The Ocean World of Jacques Cousteau

Instinct and Intelligence

The Ocean World of Jacques Cousteau

Volume 8

Instinct and Intelligence

THE DANBURY PRESS

Most species in the sea are programmed at birth to their way of life. Tightly schooled **sweepers** *remain by day in reef overhang while Australian chub swim in open water above.*

The Danbury Press
A Division of Grolier Enterprises Inc.

Publisher: Robert B. Clarke

Production Supervision: William Frampton

Published by Harry N. Abrams, Inc.

Published exclusively in Canada by
Prentice-Hall of Canada, Ltd.

Revised edition—1975

Project Director: Steven Schepp

Managing Editor: Ruth Dugan
Assistant Managing Editor: Christine Names
Senior Editors: David Schulz
                Richard Vahan
Assistant Editor: Jill Fairchild

Art Director and Designer: Gail Ash

Assistant to the Art Director: Martina Franz
Illustrations Editor: Howard Koslow

Production Manager: Bernard Kass

Science Consultant: Richard C. Murphy

Creative Consultant: Milton Charles

Printed in the United States of America

234567899876

LIBRARY OF CONGRESS CATALOGING
    IN PUBLICATION DATA

Cousteau, Jacques Yves.
    Instinct and intelligence.

    (His The ocean world of Jacques Cousteau;
v. 8)
    1. Marine fauna—Behavior. I. Title.
QL122.C636   591.5   74-23054
ISBN 0-8109-0582-5

Contents

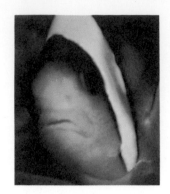

The complex mechanisms of living creatures react to stimulation from the environment. These reactions satisfy four basic needs: food, survival, procreation, and ownership. Combinations of these essential drives produce behavior that can be inherited, or learned by trial-and-error processes during life. Learning abilities pave the way for intelligence which, in turn, is indispensable to the

One of the basic motivations, perhaps the most apparent in day-to-day living, is procurement of food— is the rule in the liquid jungle, and to fulfill this essential drive, animals have developed a magnificent array of responses. The variety of feeding methods appropriate to each species has produced a broad spectrum of habits.

Feeding is only part of the endless battle illustrating an overall . Animals must be constantly on the alert for danger, whether from nature or man. The natural hazards include predators and poisonous species, fights with others, such catastrophic events as earthquakes and severe surface storms, and man's pollution.

The struggle for survival, the need to eat, and all other individual acts of behavior would be meaningless if a species were to die out. Thus the admonition takes on a special meaning. Reproductive behavior involves fertilization as well as colorful and even bizarre rituals which surround sexual activity. There are myriad ways in which fertilization does take place, sometimes on a chance basis, other times with precise accuracy.

Once launched, life must continue, and often a creature is defined by its possessions. may be essential for living since it may determine living space or breeding places. It may also convey a social status. Ownership necessarily leads to combat between members of the same species which serves to determine who is best equipped or best suited to propagate a species.

The drives behind behavior are constantly overlapping and conflicting. These drives and urges are not unrelated, but their combinations produce defining the life-style of a species. Exactly how, why, and which drives predominate at a given time or in a given place, and the

manifestations of these drives, help to distinguish one species from another, to give each its character.

The whole inventory of behavior that a species possesses adds up to constant adaptations to the environment. CHANGE THE WORLD OR CHANGE YOURSELF (Chapter VI) is the rule of the life game. Animals must adjust to sunlight, temperature, salinity, current, shelter, availability and type of food supply. It is only man, though still subject to the forces of nature, who has been able to change the environment rather than adjust himself to it.

The behavior of animals would be difficult to execute if there were no COMMUNICATIONS IN THE SEA (Chapter VII). This requires signaling of intentions, whether friendly or aggressive, useful or dangerous, harmless or predacious. It is more important to signal members of the same species as it is to communicate with outsiders.

Much of the response behavior of animals involves INBORN SKILLS (Chapter VIII). These are the genetic gifts of parents to offspring, the instinctive reactions involved in everyday living without which no animal could function, for painstaking learning could be fatal. However, the more instinctive skills a creature has, the less able it is to cope with massive changes in its environment.

There is in every animal an inborn disposition to learn, and each species has a different capacity for learning. The young of many ocean creatures are thus committed to LEARNING BY INSTINCT (Chapter IX) those skills which will prepare them for the rest of their life.

The ability to learn is a necessary condition for INTELLIGENCE AND THE BIRTH OF CONSCIOUSNESS (Chapter X), but it is not sufficient. Learning from past mistakes is the most effective way of handling unusual threats, whether they come from a predator or the environment. Learning is stored in memory, and intelligence is the faculty for establishing a relation between cause and effect and being able to associate facts that seem unrelated. With that ability, the creature can invent solutions to entirely new problems. Orcas seem to have those capabilities; consciousness is about to develop.

The chemistry that led to earth's first living cells brought about refinements from simple hungry creatures to savage warriors. All will yield to higher intellect and the STRUGGLE FOR PEACE.

Introduction: Birth of Consciousness

Among the denizens of the sea the gladiator orca is about the ultimate in beauty and efficiency. The orca is the marine mammal that inspired fear in sailors and whalers and was nicknamed the "killer whale." Herbert Ponting, photographer of Scott's expedition to Antarctica (1910-1912), reported that he was standing on a large ice floe when eight big orcas, each about 25 feet long, broke through the ice and tried to throw him into the water. Captain Scott himself reported in his journal: "One after the other their huge heads shot vertically into the air through the cracks they made. . . . The fact that they could display such deliberate cunning, that they were able to break ice at least two and one-half feet thick, and they could act in unison was a revelation to us. . . . They are endowed with singular intelligence and in the future we shall treat that intelligence with every respect." Ponting, who probably had been mistaken for a seal by the orcas, spoke of them (understandably with resentment) in such inappropriate words as, "Wolves of the sea . . . their spouts had a strong fishy smell . . . little piglike eyes . . . devils of the sea. . . ."

A recent, very reliable report relates that orcas killed a sea lion, played with it, threw it in the air back and forth like a volleyball. So orcas are obviously carnivorous, and they display ingenuity in procuring their food. They often play. But why should we emotionally judge their manners as "vicious" or even judge them at all? Marine animals have to be observed in their element, if possible in freedom, if we are ever to understand their behavior. This kind of *in situ* study was totally impossible in the past and is still extremely difficult today.

In the North Pacific our divers witnessed orcas being captured to be sold to marine parks all over the world in a new kind of slave trade. When surrounded by nets, the powerful rulers of the sea quickly understood that there was no escape. They were kept for a while in semi-captivity in a bay closed by nets, where our divers repeatedly went into the water with them. From the very start they were friendly and eager for attention and caresses from the divers. Threats were only expressed if we favored one of them and the others felt neglected. Soon they were sent away to become trained clowns.

Day after day at sea we watch orcas and wonder at the complexity of their individual social manners. They eat mainly fish and squid, though they also prey on seals and dolphins. Their family ties are strong and lasting. The head of the family is totally committed to at least two generations of offspring and the roving group constantly communicates extremely complex and even abstract information which could not be transmitted unless they had some form of structured language. On land no animal other than man himself displays such—let us say —intelligence. The long series of chances taken along the line of evolution that was necessary to turn out end products like orcas (and man) staggers the imagination.

The roots of behavioral studies lie in the automatic responses of organisms to simple stimulation (heat or light) from their surroundings. The way an animal fulfills basic motivations constitutes its behavior. And though the essential drives are only four, they can form an infinite number of combinations, just as carbon, hydrogen, oxygen, and nitrogen combine to form the highly intricate organic molecules which are the building blocks of life.

When behavior is perfectly accomplished without preliminary learning, it is innate and instinctive—all the responses to stimulation from the outside world have in this case been incorporated in the organism at conception.

But this definition is theoretical. In fact every animal, even if it is unicellular, has some ability to learn. This ability activates to varying degrees, psychological functions including memory and intelligence, although the latter plays an important role only in superior animals. What we call intelligence today is much more than the ability to learn. Intelligence includes the selective storage of facts in a memory, establishing relations between causes and effects, and the faculty to scan the memory laterally to associate apparently unrelated facts.

Evolution thus appears as progress in psychological aptitudes. From the coral polyp and the sea star to the octopus and the orca, as the share of innate behavior decreases so does the predictability of behavior. The central nervous system increases in complexity as intelligence and learning increase. The parallel progress of brain and intelligence has been followed and checked in animals up to the primates. The gigantic leap between chimpanzee and man is difficult to understand, and it coincides with the all-important birth of consciousness. The brain of a primate is much smaller than that of a man and there had been no intermediate animal to study beyond the chimpanzee until now. Perhaps the answer will come from the sea. The central nervous systems and brains of nearly all toothed whales (porpoises, dolphins, pilot whales, sperm whales, and orcas) are more developed than those of the apes and some are comparable to ours. Our closest relatives are not on land but in the oceans.

Jacques-Yves Cousteau

Chapter I. Food First

The most basic need for the survival of a creature is food. Time spent devoted to the quest for food affects the time remaining for other activities. The more primitive the animal, the more time feeding activity occupies its life and consequently the more affected are all other types of the animal's behavior.

"The basic necessity of eating influences the development of the brain."

Marine mammals, exactly as terrestrial mammals, grow during about one-quarter of their lives, in which period their food requirements are very high; when they reach adult size, hormones stop the growing process and food is only required to offset heat loss and sustain propulsion or reproductive activities. Almost all other marine creatures, including fish, carry on growing as long as they live and as long as they can find enough food. If they stop eating, they stop growing, and they grow again when food is again available. This flexibility in the fuel requirements of most marine cold-blooded creatures is obviously extremely favorable for the survival of both the species and the individual.

Eating is a basic necessity. It can be compared with shoveling coal into a furnace. In addition, eating influences the development of the brain. Superior animals develop ingenious tricks to fool or surprise their victims, thus increasing their hunting efficiency and leaving an increased amount of available time. They then develop all those behavioral activities not necessarily connected with survival: singing, playing, or inventing complicated social rules.

To find food, some animals learn how to use tools, while others hide. The means of ob-taining food becomes characteristic of the animal. How an animal recognizes what is nourishing or poisoning is inborn in some and learned in others. Many depend on a combination of inherited and learned traits for their food-catching ability. How an animal gathers food is based on behavior that depends on its physical equipment, and conversely, on physical equipment which has developed as a result of its behavior.

For example, some anglerfish have developed an extension of the dorsal fin at the tip of which is a fleshy protuberance resembling a food source for some passing fish. The anglerfish knows that to deceive his prey he must lie unflinchingly for the precise moment when the quarry grabs for the bait. With a short lunge and a big gulp, the victim is consumed. This fishing method of the anglerfish is instinctive. We can demonstrate this if we hatch an anglerfish in an aquarium where it cannot learn by trial and error. When it reaches the appropriate stage of development, the anglerfish will use its angling equipment in exactly the same manner as others of its species in the sea.

Some animals show a greater capacity to learn new behaviors, while others depend more heavily on instinct. Those that learn have the capacity to adapt quickly to new situations but must undergo a vulnerable period while learning. The animals that are "preprogrammed" emerge competent to face life but unforeseen factors can be disastrous to any that cannot learn to change.

Wobblegong shark. Presence of small fish near this shark does not necessarily trigger feeding. Particular conditions, such as time of day, and specific stimuli, such as color, smell, or motion, are necessary to generate hunger.

A Feast of Food and Sex

The life story of the adult ochre sea star (*Pisaster ochraceus*) is nothing but food and sex. The feeding behavior of this starfish is ingeniously adapted to allow maximum growth and reproductive potential for an animal with a limited amount of space within its body. For one portion of the year it eats and grows. During the rest of the year it prepares to spawn, using stored nutrients to develop gonads. These two activities follow each other as night follows day.

The ochre sea star is the most common starfish on the Pacific coast. It lives on rocky bottoms from Alaska to Baja California, indicating its adaptability to a wide range of water temperatures—from 40° F. in the north to 75° F. in Mexican waters during the summer months.

The ochre sea star, like others of its class, preys largely on shellfish. It finds prey mostly by "touch and taste," the senses used in selection of food. Frequently the prey are mussels, anchored in beds of thousands along the intertidal and subtidal zones. The

sea star feeds mostly during daylight hours at high tide by everting its stomach and insinuating it between the shells of the bivalve molluscs. With the stomach inside, it secretes digestive juices which dissolve the animal within its own shells. The sea star absorbs the nutrients through its stomach walls and then retracts the stomach. Having eaten, the sea star moves on to hunch over another victim. In summer, when waters are warm and food is plentiful, it eats voraciously. Its metabolic rate is high, but because it consumes so much food, it grows fast. The digestive gland waxes fat and gonads shrink. The star grows in a spurt during this season of eating.

During the winter months, when waters grow colder and less food is available, the sea star's metabolism slows as do its movements and its growth. Now the digestive gland shrinks and the gonads swell in size in preparation for the activities of spring spawning, filling the space formerly occupied by the digestive gland. When they are not feeding, the stars clump together to rest in subtidal waters.

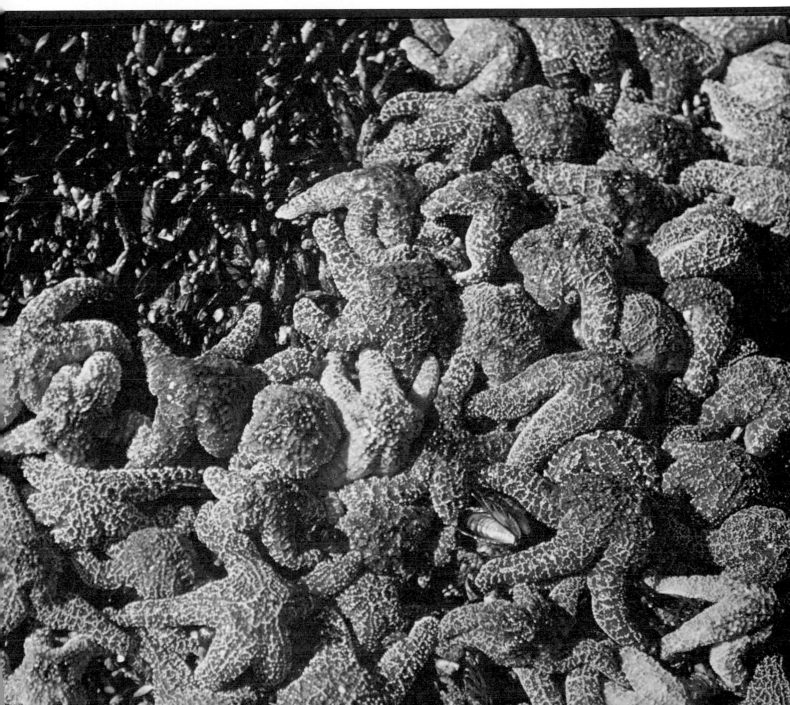

Herring gull. After landing, a gull searches for a clam it has dropped on a rock. If the shell has cracked open, the bird will eat the clam flesh.

Clamcrackers

Gulls are among a fairly large number of animals capable of using "tools." They often pick clams and mussels off a beach and carry them aloft to drop them on the hard surfaces of rocky beaches, rock outcroppings, flat roofs, or seaside roadways. Then they swoop down to pick the meat out of the smashed shells with their bills. If a clam does not crack, a gull again flies with it and drops it until the shell breaks open. The hard surfaces onto which the gulls drop the clams are, in effect, like anvils, tools they use to make prey of animals whose defensive equipment they otherwise could not breach.

Blue-footed booby. *Fishing birds of the tropic seas, the boobies are often despoiled in flight of their catch by frigate birds.*

Frigate bird. *The graceful flyers, with superior speed and strength, sometimes peck their victims severely. This white-headed individual is a juvenile.*

Social Parasites

Frigate birds of the tropics are considered behavioral parasites. They circle high over the ocean where other seabirds are fishing. When the frigate birds see the others catch fish, they swoop down and harass them until the fresh fish are regurgitated. The frigate birds, with great proficiency, deftly catch the fish before they fall back into the sea.

Common victims of frigates are the booby birds that dive-bomb into the sea's surface to capture fish they see from aloft.

Tropic birds, the strong-flying birds of the open ocean, are also victims of frigates.

Long-Nosed Feeders

All the species of fish that are in existence today have been able to make physical as well as behavioral adaptations for life in their habitats. One such group of fish are those with long jaws that enable them to pick food from narrow crevices in the coral reefs they live on. Most have adapted further by developing laterally compressed bodies. Some have tiny, sharp teeth for nipping bits of algae off rocks. Others simply suck up detritus that has accumulated.

Along with these physical adaptations are the behavioral ones that show up as preferences for feeding in the cracks on the reefs. Many of these fish have become so set in their feeding behavior that they feed only in the manner for which they are physically adapted. The long-nosed butterflyfish and the birdfish, for example, feed only by poking their long jaws into the tight spots.

When some of these fish are held in aquariums, they must be given food in this manner or they will not eat. To overcome the problem, aquarists have learned to grind clam meat or some other appropriate food and stick it on rocks. When the food has dried enough to stick, the rocks are returned to the aquariums. The long-nosed butterflyfish come to these feeding stations and peck at the food the way they would on their native reefs.

Some fish have the same physical adaptation but their behavioral pattern is different. The trumpetfish has a tiny mouth at the end of its long jaws. It hovers out of sight, close above some innocuous species like the parrotfish. When small fish come to dine on scraps left by the parrotfish, the trumpetfish picks them off. Other times the trumpetfish drifts in a vertical posture matching nearby branches of coral until it is close enough to snap up unsuspecting prey.

The pipefish and related seahorse also have tiny tubelike mouths and long jaws. Weak swimmers, these animals are living vacuum cleaners. They sip plankton off blades of sea grasses in their habitat and with great speed and accuracy suck in any smaller creatures that come close.

All these feeding behaviors appear to be instinctive, based on the physical structures the fish have inherited.

Long-snouted butterflyfish. *The long jaws of butterflyfish, at the right, enable and oblige them to probe for food in narrow crevices. Notice the deceptive eyespots on the butterflyfish.*

Trumpetfish. *The trumpetfish, below, also has a long nose but not for picking prey from coral reefs. With its elongated body it can hover unnoticed near a gorgonian or parrotfish and snap up small reef fish.*

Long-nosed butterflyfish. *The delicate beak of this butterflyfish, at bottom of left hand page, is a physical heritage that determines and restricts the feeding behavior of this creature.*

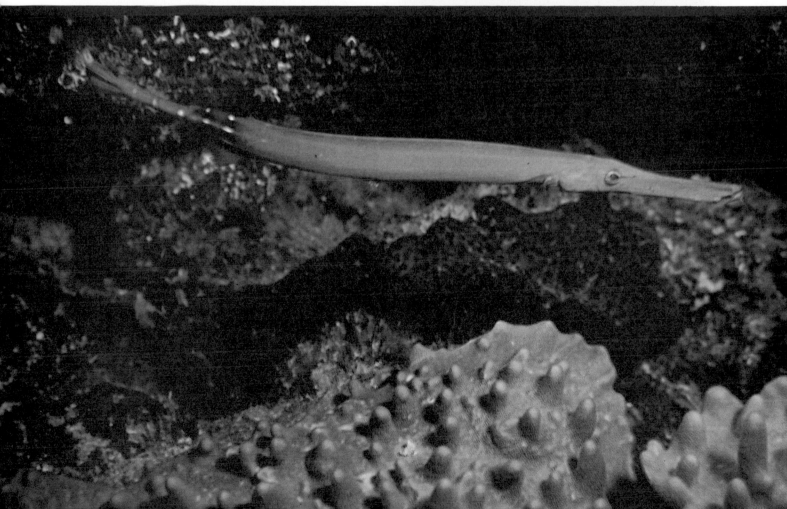

Feeding on One Another

The California scorpionfish and the two-spot octopus demonstrate an unusual predator-prey relationship—each feeds on the other. Adult octopods feed on scorpionfish, primarily the young and inexperienced. Adult scorpionfish prey on juvenile octopods.

The scorpionfish, lacking a swim bladder, is a bottom dweller and shares its habitat with the octopus among the rocks, crevices, and caves providing cover on the ocean floor off the coast of California. Since the two species live in such close proximity, encounters between them are probably frequent. But the scorpionfish can apparently distinguish between predatory behavior in an octopus and other kinds of behavior.

When a scorpionfish is approached by a larger creature, it usually adopts a defensive

The California scorpionfish, above, a bottom dweller, often preys on the small and young octopods which live in close proximity.

Two-spot octopus. Right, an octopus will feed on young scorpionfish, somehow managing to avoid harm from the venomous spines of its prey.

posture with the venomous spines on its fins standing erect and directed toward the intruder. But when the intruder is an octopus moving stealthily with its head held high and arms curled underneath—predatory behavior—the scorpionfish tries to evade it rather than standing fast in its defense.

When the octopus does capture a scorpionfish, it attaches at least two of its arms to the fish's body and pulls the victim toward its horny beak, enveloping it with remaining arms and web. Although the fish's spines are erect, the octopus is unaffected, probably

18

because of its pliability. The octopus apparently does not consume the entire fish; in both aquarium and undersea observation, the octopus left behind the trunk skeleton, head, and fins of the scorpionfish.

Those scorpionfish which fall prey to octopods are usually the smaller and younger ones, those least well equipped and experienced to escape capture. Similarly, in this reversible predator-prey relationship, it is the smaller, younger octopods that adult scorpionfish prey on, probably due to that same lack of strength and experience in avoiding capture.

The octopus and scorpionfish do not, obviously, spend all of their time preying on each other, being part of a feeding web rather than a food chain. They compete for crustaceans, which form a major portion of the diet of both animals.

The Mediterranean octopus, crayfish, and moray eel also prey on each other and constitute an even more fascinating gastronomic triangle.

The Reef Demolishers

Parrotfish are basically vegetarian, especially the smallest species of the family. But their unique feeding behavior turns the larger ones into omnivorous creatures and efficient demolition machines. With fused teeth that form a beaklike jaw, the parrotfish break off chunks of coral rock to get at the algae that grows on coral and within the living coral polyp. Then, with the teeth and paired bones deep in its throat it grinds the rock into fine coral sand and extracts the organic matter from it for food. It expels the sand and becomes the major contributor to the layer of coral sand on the sea floor on and near the coral reefs. A single ten-pound parrotfish may destroy and transform into sand one ton of reef a year. When they have thoroughly grazed a reef, they move to the grass beds surrounding the reef to forage.

Parrotfish use another method to find food. They lie on one side and, with their fins, fan rocks partially buried in sand. After feeding on organisms attached to the newly exposed rock, they may repeat the fanning procedure to lay bare additional portions of rock and more food. Parrotfish can excavate holes a foot deep in this manner.

Frequently, other reef fish follow feeding parrotfish, picking up scraps of food it dislodged while fanning the sand. Assuming these followers are not competing with the parrotfish for food, they have a commensal relationship.

Parrotfish. *In the top picture on the opposite page, the parrotfish is moving pieces of coral rubble in order to get to the food underneath. In the bottom photo, the fish is fanning the area with its fins in order to expose more food. Below, a sweetlips fish has moved into the feeding area cleared by the parrotfish which has had its fill and left.*

Nudibranch. *The nudibranch* Dendronotus *extends its fleshy white lips in preparing to take a bite of the tentacle of the mud anemone* Cerianthus.

Defense. *The anemone begins to curl its tentacles to protect them from the nudibranch. If they become tightly curled, the nudibranch may depart.*

Parasite or Predator?

An unusual variation on the predator-prey relationship in food procurement is demonstrated by the nudibranch *Dendronotus* and the mud anemone *Cerianthus*. Even though the nudibranch feeds on the *Cerianthus* tentacles, it does not, or perhaps cannot,

destroy its prey. It is rare that different species of similar size have a relationship that can be said to approach parasitism.

The tube of the cerianthid rises about six to eight inches above the substrate, so the nudibranch must climb to reach the tentacles. Don R. Wobber, a diving scientist, nudibranch could not get hold of, and the discouraged nudibranch crawled back down the tube to feed elsewhere.

The larger nudibranchs are often pulled into the cerianthid's tube when its tentacles are retracted. This occurs after one of the anemone's tentacles has touched one of the sensitive crown papillae of the nudibranch. This contact stimulates the nudibranch so that it raises the front part of its body away from the tube, arching its head back while at the same time baring large, fleshy lips and stretching its stubby body into an elongated shape. From this stretched position, *Dendronotus* snaps upward and catches one or more tentacles in its mouth. The nudibranch then gives a violent tug on the tentacle and *Cerianthus* reacts, withdrawing into its tube and pulling the nudibranch along, where it remains, feeding for anywhere from 20 minutes to two and one-half hours without any apparent ill effects.

This feeding behavior in the nudibranch is so ingrained that even when the papillae are stimulated by a tentacle artificially removed from a cerianthid, the nudibranch goes into the arching, biting, lunging reaction. And it is only a *Cerianthus* tentacle that provokes the reaction, for *Dendronotus* has been shown to be vulnerable to the stinging cells of other species of anemones.

However long the nudibranch feeds on *Cerianthus* tentacles, the damage is always minor. Why the nudibranch stops feeding, why it does not exhaust its food supply by killing its prey is still a mystery. Perhaps the nudibranch is hindered by a limited digestive capacity or by the low oxygen level inside the tube. Or it may be stopped by a mucus emission or a nematocyst threshold. Whatever the reason, *Cerianthus* begins to regenerate its tentacles as soon as *Dendronotus* leaves.

Heading for a meal. *The nudibranch has bitten the anemone's tentacle and the latter reacts by retracting its tentacles, pulling the nudibranch along.*

observed two methods of feeding by the nudibranch in Monterey Bay, California. Generally, smaller nudibranchs feed on *Cerianthus* tentacles, which are either partially or totally withdrawn. In the latter instance, the nudibranch entered the tube to feed. In some cases, however, the cerianthid curled its tentacles into tight "pigtails" which the

Chapter II. Will to Live

Defense, survival, the will to live, self-preservation—whatever it is called, it is one of the four basic motivations of life, along with drives for food, sex, and territory. Lacking the will to live, an animal might not employ any defense mechanism and would quickly become part of some other animal's drive for food. Thus, behavior and defense are inseparable.

The instinctive reactions involved in defensive behavior are passed genetically from one generation to the next. The "will to live" displayed by individual creatures integrates into the "need to live" for the species they belong to. The individuality of a species' defense, then, is a reflection of its unique genetic characteristics. Other animals may acquire behavioral defenses through learning by example from their parents or as a result of individual experience.

> "Somewhere, between inborn and learned defensive behavior, lies the dim origin of judgment ...and maybe intelligence."

The defense mechanisms take a wide range of forms; some may scare off an attacker or serve notice that taking a bite is unwise. The defense; or it may be a more complicated bright coloration calling attention to special defense; or it may be a more complicated series of maneuvers, some of which may be inborn and some of which may be learned. No prey has a perfect defense against all predators or it would no longer be a prey. As the defense improves, so must the efficiency of the attacker. Each is able to employ tricks to achieve the goal of survival, and the result is a delicate balance of nature.

Defensive success depends on rapid detection of threats and an appropriate response. The seclusive goby is so wary that it disappears at the slightest hint of a threat—sometimes even when a cloud passes overhead.

Danger stimuli may come from another animal that is not a constant threat but that becomes aggressive in certain situations or at specific times. Predator and prey have been observed living peacefully in close proximity, but the prey must be constantly alert for a change in attitude or behavior.

The defense mechanism may affect the animal's basic behavior: poisonous fish lazily swim wherever they wish, vulnerable fish group into schools to try and avoid predators, others have such a strong urge to retreat into their coral shelter that they remain there even when their home has been taken out of the water.

Animals know instinctively what defensive equipment they have and what abilities are at their disposal. These abilities may be inherited, or they may have to be developed through trial and error. Upon recognizing a threat, the animal must be able to know when and if to flee, whether to advance and challenge, or perhaps to stop and remain motionless until the threat passes. Somewhere between the inherited automatic reactions of certain animals to the stimuli of their environment, and the learned behavior of other animals, lies the dim origin of judgment, discrimination, correlation . . . and maybe intelligence.

Halfbeaks, as well as closely related surface fish like the flyingfish, often jump out of the water to avoid predators from below. At other times, they will skim across the surface of the water with only the lower lobe of their wagging tails submerged.

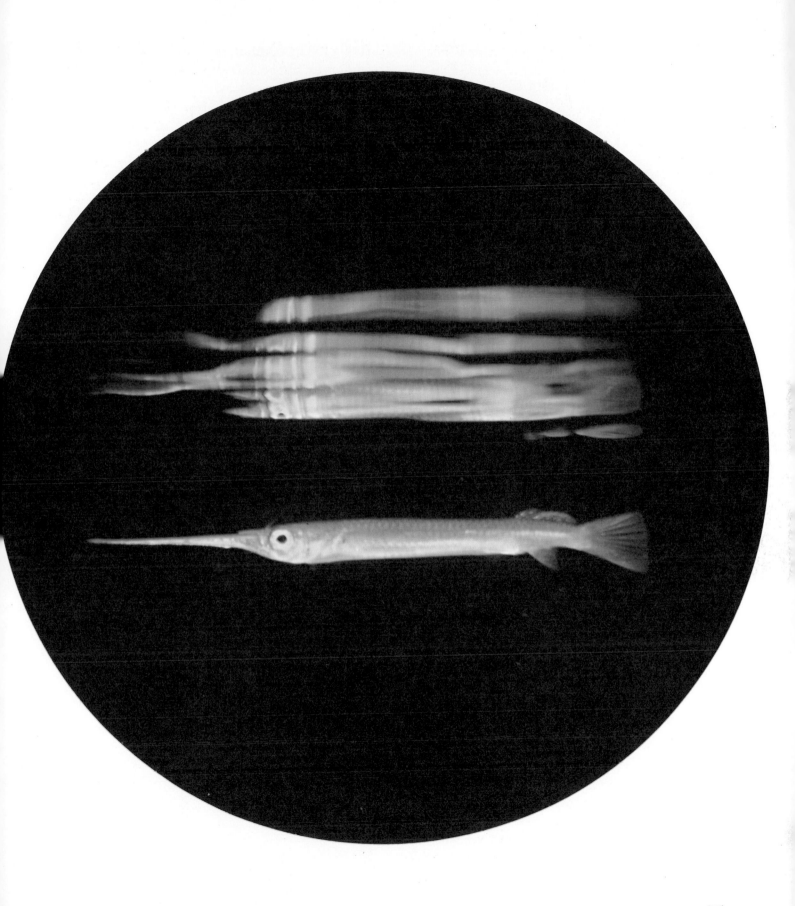

Adapting to Water

Marine iguanas are found only in the Galapagos Archipelago, 600 miles west of Ecuador in the Pacific Ocean. Their closest relatives are land dwellers in South and Central America, so their ocean-oriented amphibious behavior is unique among iguanas.

The Galapagos are tips of volcanic mountains, which protrude above the surface of the equatorial Pacific Ocean and are composed largely of hardened lava which provides a harsh environment with sparse vegetation. But there are also no natural terrestrial predators for the adult marine iguana. The eggs, laid underground by the females, are safe from mockingbirds and gulls; herons and hawks very rarely have the opportunity to prey upon a hatchling.

The adult marine iguana is aided in its struggle for survival by an ability to dive, which is better than most land animals. We have recorded dives to depths of 90 feet, during which time the iguana's heartbeat slows from a normal 45 or 50 beats a minute to four or five a minute.

The iguana, which can stop its heart for up to three minutes while submerged, shuts off circulation to its muscles, and its blood goes back and forth only between the heart and the brain.

This master diver expels air from its lungs to provide negative buoyancy and help in the descent. It can stay submerged for long periods of time, ranging up to an hour. The iguana has no gills and cannot take oxygen from the water, so it borrows oxygen from tissues of the body, which must be replenished after returning to the surface. Underwater the iguana's chest becomes concave from pressure, and it must rise to the surface by activating its flat swimming tail. Having recently evolved from being exclu-

sively a land animal, the iguana is not a good swimmer, but selective pressure may eventually produce iguanas with faster capabilities. The return to land or a reef also enables the cold-blooded animal to bask in the sun and raise its body temperature after the dive.

When there are adversaries in the water, such as a shark or a playful sea lion, the iguana can escape to the volcanic rock where its long claws allow it to hold onto the rocks tenaciously. Like other iguanas, the marine iguana displays such behavioral characteristics as territorial defense, rapid head-

nodding during courtship, and the digging of burrows to lay eggs. But during most of the year, the marine iguana is unusually gregarious, and even during the egg-laying season, the nesting beaches are communal. The territorial defense, both among males and females, is only evident during the breeding season, probably as a result of the limited soil, sand beaches, and nesting sites that are available.

During the conflicts associated with territorial defense, an unusual behavior has been observed which is probably of great survival value. A defeated individual will prostrate himself in front of his opponent as a submissive gesture. This behavior inhibits aggression in the victor and reduces the chances of physical harm coming to the loser. Submissive behavior such as this is seen throughout the animal kingdom, and even man uses a white flag or raised hands as a symbol of his submission, saying in effect, "I give up."

Marine iguanas. *The amphibious behavior of the marine iguanas of the Galápagos Islands enables them to cope with the harsh environment and sparse vegetation on the volcanic archipelago.*

Borrowing a Defense

The nudibranch is a shell-less snail. The word *nudibranch* literally means "naked gill," and thus the animal is deprived of the protective retreat that so many of its relatives possess. The eolid nudibranch is also very brightly colored and this would seem to be an open invitation to predators.

Almost all eolid nudibranchs have fingerlike projections called cerata, which vary with the type of nudibranch and may be translucent white and gently rounded. In *Den-*

dronotus the cerata may be branched like a leafless tree. Cerata probably fulfill a protective function, compensating for the lack of a shell.

Very few animals feed on nudibranchs. When handled, some shell-less snails eject a colored fluid that may offer some defense; others give off an odor that is unpleasant to man and perhaps to many ocean creatures as well. Ordinarily a fish will not touch a nudibranch, but under experimental conditions a hungry fish, which is accustomed to being fed bits of food released into an

aquarium, will snap at a nudibranch that is dropped into the water in the same manner. But as soon as the fish's mouth touches the nudibranch, it is ejected and the fish begins gaping and snapping its jaws, indicating its displeasure.

This reaction may well be caused by the eolid nudibranch's cerata, which are known at times to contain stinging cells. Although these cells are absent at some times they are present at others, and together with bad smell and taste, they are probably a source of protection. What is remarkable is that these stinging cells are not manufactured by the nudibranch itself but are derived from hydroids recently preyed upon. The nudibranch is not only unaffected by them but can consume them without breaking them down in its digestive system where they are coated by a protective mucus, and then pass them on to the cerata, where they become part of a truly borrowed protective system.

Eolid nudibranch. Reacting to prodding, opposite page, the nudibranch erects its poisonous cerata and then, below, directs them toward the aggressor.

Massing of King Crabs

King crabs are large long-legged crustaceans commercially fished along the coasts of Alaska in the north and Patagonia in the south. When young ones accumulate in a small area, they often pack, holding on to each other to form balls or pods. A single pod may contain thousands of animals, all about the same age and of both sexes. The podding activity, which is especially frequent during the first three years of the crabs' lives, continues the year round.

One small pod, containing about a hundred tiny crabs each less than one inch across, was tracked and found to move nearly 900 feet across the ocean floor in the course of two and one-half months. During its wanderings, the small pod encountered a much larger grouping containing about 6000 crabs. The two pods then merged.

The podding activity offers the young crabs some protection from predators. The individual crabs each have a tough shell, and when large numbers of them get together, their shells make a nearly impenetrable barrier for attackers. Also, the mass may appear to be a single large animal to predators, possibly deterring them from attacking the pod.

The crabs do not remain with a pod continually after joining it. When they come and go to feed, they are open to attack from such predators as sculpins, but when a sculpin attacks a pod it usually rushes into the mass, butting it viciously but usually failing to dislodge any individuals. This confirms the notion that podding for young crabs is a protective adaptation as schooling is for vulnerable fish.

Alaskan king crab. Several thousand juvenile Alaskan king crabs may be found in a large mass like this, called a pod, which offers protection against most predators.

Swimming Anemone

When confronted by a predator, the sea anemone, *Stomphia*, can swim. This is a very complex behavior for an animal with such a primitive nerve network.

Stomphia is a cold-water species from the Pacific Northwest which can detach from its substrate 50 times faster than most other sea anemones to swim with wildly gyrating motions away from its antagonist.

Sea anemones are generally sedentary, sessile creatures whose propulsive movements are so slow the human eye can barely perceive them. They depend largely on the stinging nematocysts in their tentacles for procurement of food as well as for protection from the few predators they have.

Stomphia is one of the few species of anemone which has additional defenses. Its pred-ators include two kinds of sea stars and one species of nudibranch that seem to be indifferent to their nematocysts. When a creature approaches, *Stomphia* is capable of differentiating between a predator and a nonpredator almost certainly by chemical means—a fine distinction for such a primitive network of nerves. This is well in line with modern neurophysiology which has established how incredibly complex and versatile one single neuron can be.

Stomphia doesn't use its swimming ability when it climbs onto the shell of the horse mussel, a subtidal mollusc, yet it is faster in this motion than most other sea anemones.

Stomphia *anemone. In an aquarium, when the nudibranch at right in photo above touches* Stomphia *anemone, the latter reacts by straightening up and retracting; in upper photo on the opposite page it releases its grip on the bottom rock, and begins swimming away, in lower photo, opposite page.*

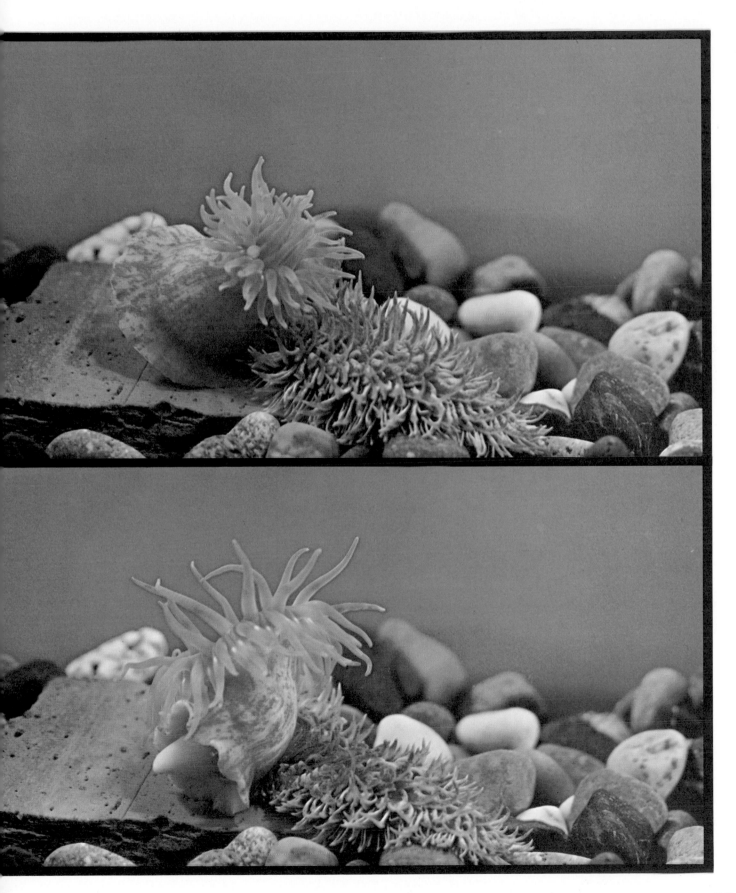

Chapter III. Go Forth and Multiply

Of the innumerable species produced by evolution over more than a billion years, the vast majority were inadequate and vanished soon after they appeared. Some flourished for a period, but became extinct when the environment changed too much. Comparatively few have survived.

In order to exist, creatures had to be adequately equipped for their daily struggles. They also had to be strongly motivated by the reproductive process. Basically chemical, this attraction is almost purely automatic for primitive species. For more complex creatures, including mammals, the drive

> "For primitive species sex is purely automatic. It becomes the reward of a lifetime for more complex creatures."

is stimulated by other factors and may involve intense pleasure and become the reward of a lifetime. The attraction of the sexes is imperative to reproduction and that attraction takes an infinite variety of subtle forms. These range from the dramatic dancing and bill fencing of the albatross to the hit-or-miss fertilization of blue mussel eggs, with countless variations in between.

The chain of events that leads to fertilization of eggs may begin with migration, the big voyage, toward a place favorable for bearing young. To reach these havens, some animals travel thousands of miles. How they know when and where to go is part of the species' memory.

At the proper time and location the individuals will pair up and proceed to the next step in reproduction—courtship. This is simply an invitation and an acceptance to mate, usually followed quickly by mating. Most often the male issues the invitation; in a few animals, however, the female is the initiator.

To attract a potential partner, animals use signals that can be seen, heard, or otherwise sensed. The male frigate bird visually signals its desire to mate by puffing up its bright-red throat pouch. Crustaceans secrete pheromones which are smelled from great distances. Only another of the same species will recognize and respond. The mating ritual allows potential mates to overcome their territorial aggressiveness. Mating is often punctuated by an attack in which the territory holder chases the partner away. Once the sex drive is satisfied, territorial aggressiveness again dominates the behavior.

Among marine fish courtship is often limited to chemical emissions or changes in color or brightness. The male dragonet also performs a ritual dance to attract a female. He swims around her many times, flashing his heightened colors and erecting his fins until she signals him to approach. Then, with his pelvic fin under hers, he lifts her and the two swim to the surface side by side. As they ascend, they shift positions until their anal fins are joined to form a groove into which they spill their eggs and milt. The fertilized eggs float to the surface and drift away to hatch or to be eaten by predators.

Those species which exist today have surmounted earlier challenges to their reproductive capabilities. But continuing environmental changes, like chemical pollution, may jeopardize future generations.

Sponges. The reproductive mechanism of sponges involves ejecting great clouds of sperm which are carried by the current toward other sponges. The emission may be stimulated by chemical communication from nearby sponges of the same species.

Chivalry Lives

Sand hoppers, or sand fleas, are minute creatures found in great numbers on sandy beaches where they congregate under wet rocks or in moist seaweed. They get their names from their jumping habit. Although sand hoppers are nominally land dwellers, they are also swimmers and, in fact, will not survive dehydration. They often dig deep burrows to avoid the drying effects of the sun and the wind.

Miniature versions of their cousin the lobster, sand hoppers have the same hard, horny outer shell that serves them in place of an internal skeleton and also as protective armor for their soft bodies. Like lobsters, sand hoppers periodically outgrow their shells and must shed them and construct new, larger ones as replacements. It is during the period after molting and before the new shell is hardened that sand hoppers are most vulnerable to predators and to the elements. Shortly after molting, in its most sensitive condition, the female lays eggs it will carry in a brood pouch until they hatch.

Here the male sand hopper steps in, as an attentive and protective suitor. Knowing instinctively when a female is about to molt and be receptive to mating, the larger male grasps the female and carries her around with him, tucked close to his abdomen. He releases her briefly while she molts, then pulls her back to his body to deposit his spermatozoa a short time before she lays her eggs. But this is not the end of the male's role; he continues to protect both eggs and female until her shell has hardened and she is able to fend for herself. For sand hoppers, at least, chivalry is not dead.

Sand hoppers. The larger male sand hopper protects its vulnerable mate and her eggs until she has formed a new shell.

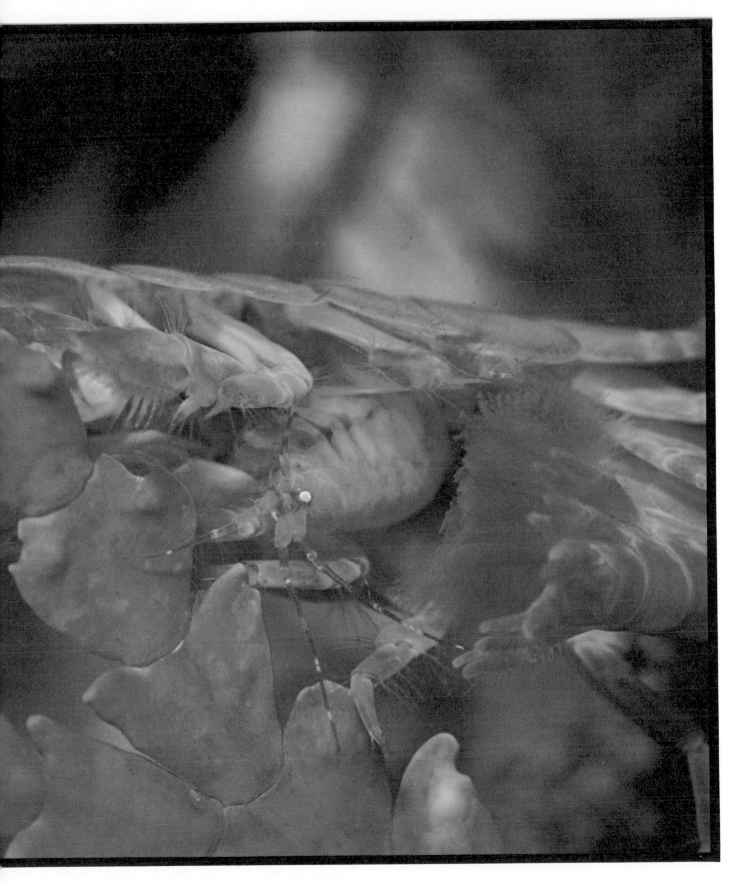

Population Control

The population of Weddell seals in Antarctica fluctuated greatly within a short period and scientists began studying their breeding habits in an effort to find out why.

The seals' breeding territories are located on ice and are determined by the presence of exit holes to the sea. These exits, which are cracks created by glacial action and tidal movements, restrict the size of a breeding colony since only a limited number of seals can use a single hole. Nonbreeders and subadults are excluded, sometimes violently,

Crabeater seal rests briefly on the ice before returning to the water, which for most of the year is warmer than the antarctic air.

from the territories until after mid-December when mating has concluded.

The population of the seals was upset when man began harvesting the animals for use in dog food. A few years after the killing started, however, there was an increase in the population, apparently due to some sort of compensatory increase in survival and reproduction rates, especially among younger seals. Their number increased to nearly

3000, the largest recorded, but four years later declined to slightly more than 2000 individuals. This decline occurred even after the killing by man had stopped.

One year the scientists found two dead seals on the ice. There had been none the previous two years. The next year five dead seals were found, and the following two years 15 and 14, respectively. The dead seals showed sufficient layers of fat to rule out starvation as a cause, and most of the dead were adults over ten years old. These dead animals, none of which appeared to be seriously wounded, were found between October and the end of December, suggesting some connection with the breeding season, since this is the time of the greatest intrasexual strife in competition for breeding territories and rights.

The conclusion, then, was that as a result of the intraspecific killing, the population began to decline again and continued to do so only to the point where there were the maximum number of individuals that the breeding territories could accommodate. This adjustment was harmful on the individual level but beneficial for the species as a whole.

Breathing hole in the ice gives the seal a water exit in which to fill its lungs after a dive of nearly an hour's duration.

Wrasses. Usually territorial, wrasses are grouped over their home reef with a few males and many females in school swarming around together.

Hit-or-Miss Spawning

The eggs of most fish are fertilized externally. A female lays the eggs in a nest or perhaps just releases them in the water leaving them to the vagaries of ocean currents. A male must deposit his milt in just the right place or the eggs will remain unfertilized. A number of species in the wrasse family spawn in a well-regulated fashion after gathering in schools over their home reefs for a few hours during the day. Periodically, a group of five to 15 males and females from a larger group bunch up tightly and swim rapidly toward the surface. When they are four to six feet above the main school of wrasses, they begin releasing eggs and milt into the water, leaving to chance whether they come together and result in fertilized eggs. Any current or tidal effects quickly scatter these gonadal products, so fertilization must take place quickly or not at all.

Separating. *Out of the larger group, a small number of males and females separate and begin to mill around more and more frantically.*

Spawning. *The smaller group of wrasses rushes upward and as females scatter eggs into the water, the several males spread their milt to fertilize them.*

Mating in aggregations is customary with those species that tend to scatter their eggs. Fish that spawn in this manner also lay large numbers of eggs to increase the chances of their fertilization and survival because eggs scattered in the sea are subject to predation far more than those laid in nests guarded by parent fish.

The same is true of other marine animals. Barnacles and oysters, for instance, are drifters in their larval stages that settle permanently in large colonies of their own kind. Blue mussels lie in beds of thousands in shallow coastal waters of America and Europe. During the spawning season when females release their eggs in the water, it serves as a chemical signal to the males to release their sperm, fertilizing the eggs as they drift by. Spawning in aggregations with large numbers of eggs is an instinctive behavior pattern which makes the fertilization of eggs almost certain.

Breeding Bulls

Before man interfered with the natural processes, either through overkilling or overprotection, sea elephant colonies were relatively small groups scattered along coasts and islands. The harem reproductive behavior was strictly enforced.

The fierce fights among breeding bull sea elephants happened to help keep their species strong. The strongest bulls were those which got to mate with the harems of females that arrived on the breeding beaches four to six weeks after the bulls.

When the bulls arrived, they vied with each other in ferocious encounters for social position. The No. 1 bull, or beachmaster, had first choice of the cow sea elephants.

But today man has exterminated sea elephants on all coastal beaches and has relegated them to a few islands where they pile up in overcrowded colonies; mating and maternal instincts have been disrupted and the animals have become promiscuous.

Sea elephant challenge. A bull (left) bellows an invitation to fight. Social status is determined by contests of strength.

Battling bulls (below) may resort to blows if neither retreats. The victor will win the right to form a breeding harem.

Reversed Sex Roles

In some species the male is more brightly colored and aggressive than the female, who incubates eggs and rears the young. In others, the male and female look and behave very much alike, even to sharing incubation duties and caring for the young. In a few species, though, the sex roles seem to be reversed. The female is brightly colored and more aggressive than the male, which cares for the eggs and the young.

The sandpiperlike phalaropes are just such animals. In courting, the female is the aggressor. Her bright plumage attracts the dull-colored male. After she lays her eggs, she leaves. The male sheds his breast feathers and the bare skin beneath becomes engorged with blood, forming brood patches. The heat of the blood in these patches enables the male to incubate the eggs over the next three weeks. After they hatch, he cares for the young for one and one-half weeks, until they can shift for themselves.

This reversal of roles can be traced partly to a physiological factor. The male hormone, androgen, which tends to make animals aggressive and brightly colored, is found in equal or greater amounts in females than in males while males have more of the female hormones. Thus, internal physiology can be just as important as external stimuli in affecting behavior.

Phalaropes. *Brightly colored females, left, fight to woo the smaller, dull-colored male. Mating pair, below, assumes a breast-to-breast courting posture.*

Chapter IV. Ownership

Since the very beginning of life on this planet, life for plants and animals has meant a perpetual fight against all things and creatures. Cold, heat, currents, storms, enemies, poisons, famine, parasites, and sickness build up a world of universal hostility with barely a few moments of peace, friendship, and love. It is success against a constant struggle that makes life the most precious of all goods.

Perpetual fighting creates an instinctive need for power, for domination. Only man can actually live the hopeless dream of universal domination; all other creatures relate their ambitions to a reasonable territory where they will be the rulers.

Thus territoriality, as private property for humans, has its origin and its justification in the basic hostility of all environments. Dolphins, having practically nothing to fear, ignore ownership or domination. Now that humans are the masters of the planet and have no enemy except themselves, private property is losing its original significance.

Territoriality assures that the species will be dispersed and that there will be enough food for each successful individual.

But living space and food are not the only reasons for defending territory. Seals defend it because of the concomitant breeding rights. Anemonefish defend their anemones in order to assure themselves of a hiding place in the future. And pilotfish keep other members of their species away from their shark not because it is a living space or necessarily provides food, but because the shark is a source of protection from predators.

Territorial drives are not equally strong at all times among all individuals of a species. While it is true that some shrimp attach themselves to nudibranchs, apparently for their lifetime, other animals are territorial only during certain times of the year or during certain stages of their lives.

> "Territoriality for animals, as private property for humans, has its origin and its justification in the basic hostility of all environments."

Territoriality, as procurement of food, self-defense, maintenance of social status, or mating challenges, may generate fighting. In the case of self-defense, the battle may be fought to the death. There is a difference between fighting among members of the same species—intraspecific aggression—and fighting with members of another species—interspecific behavior. Territoriality involves intraspecific fighting. Once a territory has been claimed, it is often defended solely by boundary displays. Fighting then occurs when an intruder crosses the border. But in other forms of aggressive behavior, fights are often ritualized, and killing is rare. In these conflicts of intimidation, each opponent tries to look larger or more fearsome than he really is.

In their tournament fights, fiddler crabs ritually wave their lone claws, then butt each other's claws. When the fighting gets serious, a fiddler crab can only grasp his opponent's claw in a very specific manner, which virtually assures little or no damage to the loser.

Home in a sponge. Large, tube-shaped sponges furnish living space for many species of gobies. Males guard eggs laid deep within the sponge and feed almost entirely on the worms that are parasites of the sponge.

A Living, Moving Territory

Many shrimp live in close association with other animals, mostly sessile creatures like sea anemones and sponges. A few shrimp live on creeping animals like sea urchins. And some, like the beautiful shrimp *Periclemenes imperator,* have been observed living on an active mollusc, in this case a free-swimming, shell-less dorid nudibranch.

Both the shrimp and the nudibranch are brightly colored, in similar shades and patterns, and only one shrimp has been found on a single nudibranch. When a second shrimp has been introduced, the first one attacks it in an attempt to drive it off, often successfully, suggesting that the shrimp considers the host nudibranch to be its own territory. In fact, in its natural habitat this shrimp has never been observed off the host nudibranch. If a shrimp is removed or shaken from its host, it frantically scurries about until an antenna or dactyl touches the nudibranch, whereupon the shrimp remounts its host.

The shrimp rides the dorsal side of the nudibranch and is apparently of no special concern to its host, even when the shrimp touches or pinches material off the surface appendages, whereas at other times, when these gill filaments or rhinophores have been touched by foreign objects, they are immediately retracted.

The shrimp benefits from its "possession" of a nudibranch, since the host wanders about in search of food and increases the variety and range of potential food sources for the shrimp. When the nudibranch settles on the bottom to rest or forage, the shrimp can feed by reaching over the side of the mollusc to pick up material from the surface on which the nudibranch is resting. In addition, the shrimp will pick up such material as lumps of mucus and detritus which it finds on the surface of its host.

The nudibranch swims by powerful undulations of its body that would seem to make it difficult for the shrimp to stay aboard. But the nudibranch starts slowly, and the shrimp is able to detect the movement and lodge itself in a groove between the fleshy mantle and muscular foot of its host.

Since the shrimp never leaves its living territory, and there is only one shrimp to a nudibranch, reproduction would also seem to be a problem. But the shrimp probably mate when two nudibranchs get together for their mutual exchange of sperm.

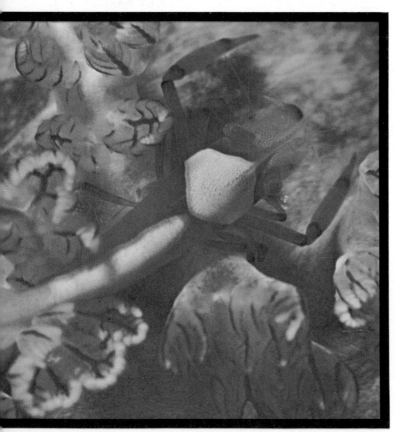

Host. *The free-swimming nudibranch* Hexabranchus marginatus, *opposite page, provides transportation and a mobile home for the beautiful shrimp* Periclemenes imperator.

Hitchhiker. *Left, this shrimp will never leave its nudibranch and will prohibit another shrimp's access to its host as though it were a territory.*

Claws as Status Symbols

Ownership of claws is a determining factor in the life of crabs, lobsters, and many other species of small and large crustaceans. Regardless of whether male or female crabs practice territoriality, it is the possession of claws that determines their status.

In most species of crab, the claws of the male are larger than those of the female, and as a result, it is the male which engages more frequently in combative activity. During periods of antagonistic behavior with other members of the same species, the order of dominance is based on body size and sex, with status signified by the size of the claws. Combative behavior is often highly ritualistic, featuring a rather monotonous display of intimidation. In the case of the fiddler crab, it was learned that the waving of its lone claw precedes a very stylized fight in which the combatants are rarely injured. The shore

crab, however, uses both claws without fa-
voring one or the other. In some cases,
though, ritualism is not enough. One observer noted that a medium-size female crab
had established dominance over a number
of both male and female crabs by an extensive display of claws. When a larger male
crab appeared on the scene and the dominating female tried to intimidate him, he immediately attacked her and ripped off one of
her claws. The beaten female retreated, but

*Defensive claws. Most crabs use their claws de-
fensively, but often this defense is symbolic: the
claws are waved to intimidate foes.*

was pursued by the group that she had previously dominated, since her status had been
greatly lowered with the loss of a claw. This
occurred in a species where the females have
large claws and probably could not have
taken place among a species where the females have smaller claws than the males.

51

Mouth-to-Mouth Defense

The jawfish is found in most warm waters of the world, with the various species living in burrows that are usually lined with bits of rock or coral for reinforcement. These carnivorous fish are "tail standers," using constant motion of the pectoral and tail fins to remain almost vertical in the water as they hover pugnaciously near their burrows. At the first sign of danger, however, their courage vanishes and they dart back into their homes, usually tail first.

Jawfish in captivity, however, display a strong territorial behavior, for when another fish enters the surface area of a burrow, it is quickly intercepted. Actual contact is usually avoided, but there are display threats where the jawfish will erect its fins, flare gill covers, and threaten to bite by opening the jaw widely. Occasionally the fight may continue to the point where jaws are locked and a pulling fight ensues, but this is rare.

Yellowhead jawfish. Territorial defense of the jawfish includes threatening with the mouth, though actual biting is rare.

Defending Mobile Homes

The hermit crab defends its territory, usually an empty snail shell in which it lives, with a highly stylized ritual, which virtually assures that neither it nor its opponent will be seriously injured.

In preparing to do battle, the hermit crab signals its intentions to its opponents in one of three ways. It may first move its legs laterally, which may be compared to a boxer's shifting his weight from one foot to the other. The hermit crab may then raise its claws—displaying its weapons—by holding them vertically, perpendicular to the bottom. The most aggressive stance, when the crab goes into its "fighter's crouch," is the extension of its claws in front of it, ready to do battle.

In addition to defending territory, male hermit crabs may "own" a female during the breeding season. In many species the male is larger than the female, and the male often grasps the edge of the shell in which a female is living and hauls it around.

While there are some hermit crabs, such as the very large *Petrochirus diogenes*, which do considerable harm to each other in fights under laboratory conditions, many other species have replaced combat by a ritualized act of shell tapping. By banging their shells, competitors in a ritualized fight have a much lower chance of damage and the loser surely has a better chance of eventual reproduction. This is not so in confrontations with members of other species. Some crabs may gang up on an intruder and kill it by rocking its shell and reaching in to grab its legs.

Hermit crab. A lateral movement of the legs (top) is the initial stage of aggressive display for a hermit crab. It may then display its claws (center) before assuming its battle posture (bottom).

Aggressive and Fearful

In a few specific instances, fur seals and sea lions can become aggressive animals, fighting for territory established during the breeding season. But intimidation displays may also occur underwater, as some divers can attest. We still have a lot to learn about undersea behavior of seals. Reliable reports are rare, but it seems that seals are more likely to act aggressively when their young are nearby. With the California sea lion, this aggression is usually expressed by extremely rapid swimming directly at the diver with an abrupt turn at the last instant to avoid collision. This aggressive rush is usually very effective in chasing divers since underwater the glass plate of a diving mask has magnification properties; the threatening aspect of the sea lion hurtling toward the diver is therefore greatly increased.

This reproduction-related aggression is influenced by the fact that females usually breed with the territory holders. Fur seals, as elephant seals, also fight for status in the social hierarchy and this rank determines which bull has access to the most females.

In the antarctic, where man's predation has had little influence on behavior, seals and fur seals have no fear of man when they rest or crawl on the ice or on shore. This is probably the reason sealers can slaughter them so easily.

Underwater, however, these seals flee at the approach of divers, demonstrating that their only natural enemies come from the sea (most probably orcas) and that their behavior is more instinctive than intelligent.

California sea lion. This sea mammal, which weighs as much as 800 pounds, has been known to threaten divers by rapidly swimming directly at them, then turning away just before impact.

In Defense of Sharks

With a shark, nothing is certain. This is a result perhaps of our ignorance rather than the shark's incoherent behavior.

Aggression in sharks has most often been assumed to be the result of hunger, with a diver as the potential meal. Attacks on struggling swimmers or downed pilots appeared to be the result of acoustic, visual, or chemical signals which triggered a feeding response in the shark. It is certainly untrue that sharks are mainly carrion eaters and attack only the disabled. There are no facts to support the notion that sharks are especially attracted to decaying meat or decomposed flesh. A hungry shark will, of course, bite at this, but it would probably bite at just about anything. A shark rarely has a very big appetite. Its energy needs are low: it swims in unobstructed water and does not suffer from cold, so it has little caloric loss. The shark's reputation for insatiable hunger would appear undeserved. But sharks attack for reasons other than sating hunger.

It is revealing to examine the shark's feeding behavior and note the similarities between it and attack behavior.

When sharks feed opportunistically on large food, such as a dead whale, they usually cir-

cle it slowly, sometimes for a considerable period of time, before making a pass. When the pass comes, the object may first be bumped with its snout, and then on subsequent passes, it is attacked. At one time people thought the shark could only attack a large floating object upside down because of its ventrally located mouth; but some sharks are able to snub up their noses and with a special system of muscles and bones evert their jaws, enabling them to take a large bite from the side of a large prey without turning over. Once the shark bites, its body and head swing violently back and forth as it uses its parallel rows of teeth to saw off food from its victim.

The shark sees well, possesses a refined sense of smell, and is extraordinarily well equipped to receive low-frequency vibrations. We think that the vibration waves generated by a fish in trouble will bring the shark, because the shark's flanks are paved with sensory cells and thus its body is an antenna for receiving infrasonic impulses from great distances.

New theories support the proposition that attacks by sharks on people are not all unpredictable feeding actions. One scientist suggests that the aggression is really a defensive reaction by the shark when provoked or pursued, especially in a confined area. Another suggests that the aggression may

be triggered when the shark's personal living space is violated.

As with many animals, aggression isn't all biting and fighting but most often is displayed as some sort of ritualized act or threat that may be followed by a fight or attack. An aggressive shark usually swims in a laterally exaggerated motion while the body sometimes gyrates in a spiral and looping manner very similar to the feeding behavior. At the same time, the posture is alerted, with a lifting of the snout, dropping of the pectoral fins, arching of the back, and lateral bending of the body. In several observations, this behavior was immediately followed by an attack on a nearby diver or object.

Observers of one incident of preattack behavior related, "The shark started a small circle just opposite us and as he came around his body started turning and twisting and rolling back and forth in the water as he swam. The whole body was being used to swim with, his head moving back and forth almost as much as his tail. . . . The shark continued this small circle and as he headed toward the reef he straightened out and began to increase speed slightly. . . . He abruptly turned and made a lightning-speed pass directly at Jim. . . . The shark narrowly missed him, went around in a sharp U of less than five-foot radius and came back directly at Jim's face. Jim had just enough time to swing around in the opposite direction and throw out his arms in defense. The shark grabbed his elbow area, gave two quick bites, and flashed away. All this happened in less than five seconds."

In another case an intense display was followed by the shark suddenly whirling and biting the Plexiglas hood above the heads of two divers in a small research submarine. The divers said the attack was of such force that it sounded like the hood of a car being slammed as hard as possible.

The aggressive display does not always result in an attack, however, as diver-scientists Richard Johnson and Don Nelson found out. Working at Eniwetok Atoll, they aggressively approached and cornered a number of sharks under controlled conditions, and in each of ten test trials they saw the ritualized movements and posture described above. The intensity of the shark's response, or threat display, was related to the degree of opportunity for escape. The most exaggerated postures occurred when the degree of confinement was greatest.

Johnson and Nelson concluded, "This behavior is probably a defensive threat, is ritualized, and may be part of the normal social encounters among sharks. This may be a motivational basis for attacks on man for reasons other than feeding."

Their mention of "social encounters" among sharks fits in with other studies which indicate that some sharks have a social order of dominance within a species and that dominance relationships may actually be established among the different species.

All in the Family

When directed at a member of the same species, aggression is generally nondestructive. Then the aggressive behavior is manifested in stylized mannerisms. In many cases the aggression may be a ritualized form of another type of behavior. For example, fish that normally feed by grazing may use their mouths in a similar way when they are engaged in aggressive activity. But such ritualization of feeding behavior is not always the case. Mouth breeders, whose delicate mouth serves two important functions, seldom bite during aggressive behavior. Instead they may circle each other and engage in a bumping contest.

Nondestructive aggression is displayed in another way by these spectacularly colored lionfish, as well as butterfly cods, turkeyfish, and zebrafish, which all belong to the scorpionfish family. Some are well camouflaged by their coloring and patterned stripes for living amidst underwater debris, but the coloration of others is probably a warning that they would not make a palatable meal. They have elongated and slender fin spines, with small and well-developed venom glands.

While engaged in antagonistic behavior with other species members, these fish threaten with aggressive movements and flared fins. But they never use their venomous spines against a member of the same species.

Each of the many spines of these fish contains a needle-shaped venom gland, which may extend for as much as three-fourths of the length of the spine. The apparent absence of a glandular duct through which the venom may be ejected means the venom is probably released through pressure on the spine, such as when the fish is bumped hard or roughly handled. Their stings are considered more dangerous than sculpins' and stingrays' and may be fatal to man.

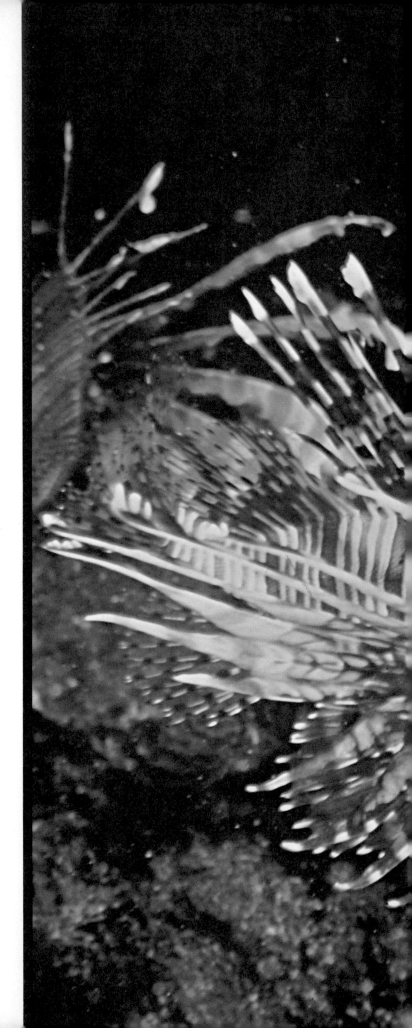

Pilotfish. A shark, a boat, or any large fish may be the attraction for pilotfish, which have been seen defending sharks as though they were territories.

Pilotfish on Defense

When two sharks come close together, their attendant pilotfish change color from their normal black vertical bars to almost solid silver-gray. This is apparently a display mechanism which signals aggressiveness. This change takes place only when fighting between pilotfish occurs and only in the presence of more than one shark, so it is presumed to be closely linked with territorial defense. In pursuing each other, the pilotfish may stray several yards from home base before returning.

Pilotfish may benefit from the shark's presence by seeking refuge from predators.

Sea anemone. *This solitary animal may use its stinging cells in an aggressive manner against a relative that settles too close.*

Fighting Anemones

There are some species of anemone that have pouches or pods, called acrorhagi, filled with stinging and poisonous cells (nematocysts) and located at the base of their tentacles. Normally anemones are solitary and passive, but unusual behavior has been ob-served among individuals of the genus *Anthopleura*. When one anemone settles too close to another of this genus, the original resident will inflate or enlarge its acrorhagi and then press them against the skin of the intruder. If the intruder does not move away, it will be disintegrated by the poison of its disagreeable relative.

Chapter V. Complex Manners

Mammal, fish, crustacean, mollusc . . . each animal is faced with the same basic needs for the continuation of life. It must succeed in the face of elements and enemies; it must obtain food, defend itself or its territory, and propagate the species. The manner in which each species performs these tasks is unique, and the combination of these behavioral patterns helps to differentiate each species from all others.

Some animals own a home, a burrow, a particular coral head, or a patch of bottom. Others wander the reefs or open sea, possessing no home and never even pausing to rear their young. Animals can be distinguished by other behavior, such as how they feed, the manner in which they reproduce, or the way in which they defend themselves or their territories. These varied activities, no matter how ordinary or bizarre they might seem to us, combine to help achieve the survival of the individual and the species.

> "The inventory of responses to the four basic behavioral drives has an unlimited number of potential combinations and patterns."

Just as each species possesses a unique set of chemical entities and systems that make it different from every other, so does its behavior differentiate one species from another. In fact, some species are very closely related genetically and could possibly produce offspring, but they remain isolated because of behavioral differences. It may be that the mating dance or vocalizations or signals of one individual fails to trigger the proper response in a potential partner, and as a result reproduction does not take place.

Behavior itself is strictly a mechanical action, a response to the four basic drives for survival, food, reproduction, and ownership. But these responses combine to produce an inventory of behavior in animals. This inventory has an unlimited number of potential combinations and patterns. Sometimes there is overlap and sometimes there is conflict.

In this chapter we will choose one example of such complexity of habits and explore how various drives combine to make up the behavioral inventory in damselfish. Even though the different species of this family are genetically similar and each has the same basic drives, each is behaviorally different and can often be identified by its actions.

Survival and feeding behaviors interact to prevent some damselfish from venturing far from home in search of food and result in a pattern of quick darts to capture a meal and retreat to safety. In some species, schooling behavior, which is of survival value, is eliminated during the mating season, and the males become territorial. Internal factors, possibly hormones, affect this change from the point of view of individual survival (that is, schooling) to one of species survival (that is, reproduction). In some cases there is even a social structure of dominance, but even this behavior is not isolated from other activities and is influenced by sex, with males being more aggressive and elevated in the social hierarchy.

Blue demoiselle. The damselfish family is widely distributed in tropical, subtropical, and temperate waters. This species, Chromis caeruleus, *is found from the Red Sea to the western Pacific.*

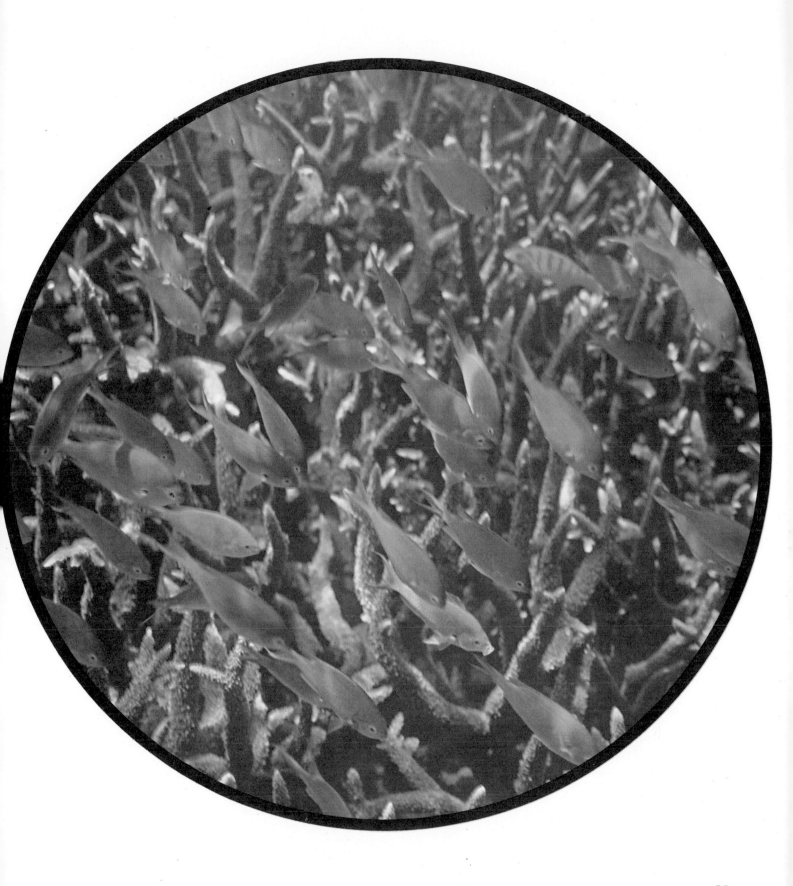

The Family of Demoiselles

The damselfish family offers a unique opportunity to look at behavior patterns in depth. Damselfish are found throughout the world, so generalizations and comparisons may readily be made based on observations of very close relatives.

Damselfish are generally most active during daylight hours and have a tendency to remain localized. Characteristically damselfish cluster around some object, such as a rock or coral head, which they use as shelter and may remain there for months at a time. This is not restricted to breeding individuals; juveniles as well as mature males and females have been observed in this pattern. Some damselfish species, however, show a preference for living among anemones rather than coral, but many of their other traits reflect typical damselfish behavior.

This resident status and diurnal activity make it easy to observe feeding habits and reproductive behavior. During night hours the fish retreat to their shelter rock or coral, with a few solitary individuals remaining quiet but alert.

Damselfish generally live in warmer ocean waters, and the food habits of the more than 200 species are far-ranging, with some being herbivores, others zooplankton feeders, and still others feeding on anything from plant and detrital material to small crustaceans and anemones.

What little nighttime activity there is, and this is more common for solitary individuals than those in large aggregations, is concen-

All around the world. Chromis *damselfish, like* C. cyanea *in the top picture, are found in Discovery Bay, Jamaica, while their close relatives in the bottom picture,* C. caeruleus, *can be seen among the Seychelles Islands in the Indian Ocean.*

trated close to the bottom when there is bright moonlight. A number of individuals of the species *Abudefduf troschelii* were taken at night under a bright moon and showed no food at all in their digestive tracts, indicating that feeding was not the primary purpose of this nighttime activity.

Perhaps one of the more remarkable feeding behaviors among damselfish occurs in *Chromis atrilobata,* which has been observed at times feeding on planktonic fish eggs, such as those of the wrasse. This is not their exclusive diet, since during large parts of the day many damsels feed on midwater organisms, mainly small crustaceans.

When spawning, the wrasses gather in large groups and swim in the same area for several hours. Periodically, a small group of perhaps five to 15 individuals of both sexes will draw themselves closely together and begin swimming almost vertically, in a rapid fashion toward the surface. After swimming approximately four or five feet, the group makes a sharp turn and reverses direction, heading back down to the larger aggregation. As the abrupt 180° turn is being made, the wrasses release their gametes into the water, where fertilization will take place.

The hovering damselfish observe this activity and converge on the ascending group to feed on the eggs when they are released. The damselfish has either learned or instinctively knows to recognize this behavior of the wrasse. Proper timing is important, for if the damselfish arrive prematurely, while the wrasses are ascending, the wrasses would return to the bottom without spawning.

The habitat. The damselfish are most active during the daylight hours when they leave their shelter to feed. Most species live among coral but there are some, as we will find out later in this chapter, that live within sea anemones.

Environmental Influences

Since the damselfish is diurnal (active in the daytime), its comportment is influenced by environmental factors, in this case light. Observers have pursued even further in an effort to determine the correlation between environmental factors, like light or water current, and behavior, like feeding or spawning. Among the damselfish genera studied in this way are *Eupomacentrus, Chromis,* and *Dascyllus,* as well as *Abudefduf.*

These damselfish display what might be called a "conflict of instincts," since they have a strong desire to stay close to the protection of their home rock or coral but must leave it in order to secure food. In the morning hours, when the urge to feed dominates, active feeding occurs. Since the food supply relatively close to shelter has not been depleted, damselfish do not have to stray very far from home. In time, though, the

On the large side. Damselfish of the genus Eupomacentrus *are generally larger and less brightly colored than other damselfish.*

amount of available food diminishes; feeding activity slows down as long as the tendency to remain near shelter is strong enough to overcome the tendency to swim into the current to get food.

Influencing these conflicting urges are two major environmental factors: light and water current. Observers have noticed that damselfish feed less rapidly on cloudy days than on bright days, so there is an apparent correlation between the intensity of the light and the amount of activity. The presence of current—which constantly replenishes the food supply—also influences behavior. Thus it is not surprising that the most active feeding is done at time of high light level and during the hours when water currents reach their full strength.

Environment and instinct conflict when the fish want to feed but there is no current bringing planktonic material close to the shelter. If there is no current, the fish must swim farther from the shelter in order to feed. Thus, since leaving shelter is done only to feed, it has been noted that the greater the distance these fish are from shelter, the greater is the feeding activity. There is also a gradation in size and distance; that is, the larger the fish, the farther they were from shelter. Or to put it another way, those fish farthest from shelter were generally the larger adults, while those closest to the shelter were the smaller juveniles.

The current does more than just bring food to the damselfish, for it provides a point of organization and orientation. When the current is absent, the fish may wander in what seems to be a random pattern without engaging in specific behavior.

Another environmental factor which influences feeding behavior is turbidity. When visibility decreases, the damselfish retreats to shelter. If the water clears for even a brief period, the fish leave their shelter to feed. But when conditions of poor visibility persist for several days, the fish gradually move farther from shelter. Apparently the hunger drive overcomes the urge for shelter.

Feeding, as noted, is the major activity of the day. No large numbers of individuals are involved in reproductive behavior at any one time. What sexual activity there is usually takes place in early morning.

The reproductive male spends little time feeding when eggs are in the nest. His time is spent in territorial and parental behavior.

Guarding the nest. This blue damselfish, like most male pomacentrids, usually performs the task of egg-guarding after spawning.

ATTRACTING THE FEMALE

CLEANING THE NEST SITE

EGG LAYING

Spawning Demoiselles

The demoiselle, or damselfish, *Chromis dispilus,* features nursery behavior and nest guarding among males during the incubation period of the eggs. This particular species is found off the coast of New Zealand; its spawning activity begins in the early morning hours but continues much later into the day, sometimes until afternoon.

Spawning occurs on bright days when the water is calm and underwater visibility is maximal. The foot-wide, oval-shaped territories are randomly scattered on smooth rocks, each defended by a lone male whose normal bluish gray to blackish green coloration had greatly intensified. An aggregation of nonmating males and gravid females swims above, about four to six feet over the territories. Occasionally a fish approaches but if it is not of the same species, it is driven off. A female laden with eggs is usually accepted; she swims to the nest site where the territory holder moves to her side, usually

DEFENSE OF TERRITORY

TERMINATION OF SPAWNING

FANNING THE EGGS

facing in the opposite direction. "Tail flashing" is the way the male attracts the female during courtship. Tail flashing involves rapid exposure of the tail fin, causing the white margin to "flash." The male and the female then engage in a quivering motion, skimming the rocky surface, during which time eggs are extruded over the surface. The whole process takes two or three seconds and may be repeated three or four times. As soon as this activity is over, the male chases the female away, his reproductive urge tem-

porarily satisfied and his territorial urge temporarily dominating his behavior. He then returns to the nest and begins to fan the eggs vigorously with rapid movements of his tail and pectoral fins. The male usually does this by leaning to one side, and the fanning is an attempt to provide maximal oxygen for the growth of the eggs. Any bits of debris, small invertebrate intruders, or even objects placed by humans on the territory are immediately carried away, or if they are large, pushed to the edge of the territory.

Male Nurses

One of the characteristics of damselfish is the nursery role played by the male. In the genus *Abudefduf,* the males care for the eggs by cleaning them and driving away most intruders. They are able to differentiate between a predator which is likely to feed on the eggs, like wrasses, and the nonpredator, and actively drive off the former. It has not yet been determined whether this differentiation among predators by the damselfish is learned behavior or instinctive.

In the *Abudefduf* males the complete reproductive sequence starts at the beginning of the season when schools made up of both males and females begin exploratory swimming along the coral reef, hunting for the location of nesting sites. The males head the school and exhibit a color change from the usual stripes to blue, a color that becomes

Sergeant majors. These fish belong to the genus Abudefduf *and are known for their ferocious nest guarding. They are somehow able to distinguish between egg-eating predators and other creatures, and are most aggressive in driving off the egg-eaters.*

more vivid as the spawning activity reaches its climax. The selection of sites is seen as a "dive" from the group by certain males, and this dive halts the activity of the entire school. The nest sites are bare or eroded places on the coral substrate, sometimes the same places that had been used in previous spawnings. It has not been determined whether these locations are recognized by intelligence or chosen instinctively. The males then engage in such preliminary activity as displaying erect dorsal fins and thus begin the establishment of individual territories. Continuous threatening, chasing, and biting aids in defining the boundaries, as the

coloration of the males becomes an intense blue. Aggressive behavior gradually diminishes once the territory has been well defined, whereupon the males begin cleaning the nesting site by nipping off debris.

Courtship involves short vertical "invitation" swims and return to the nest, with this activity directed by the males at the females passing the nesting site. During spawning the male and female swim in a circular pattern, and as the mass of eggs being deposited grows larger, the female becomes less regular in her circular swimming, which in turn stimulates aggression in the male. After a period of about six or seven minutes, the egg-laying activity ceases, and the male drives off the female. The male's spawning activity may continue, however, with five or six different females.

The male then begins egg-cleaning and nest-guarding activity. During this time coloration changes back to normal and soon after the hatching is completed, the male resumes his nonreproductive behavior.

Defending a Territory

As noted earlier, light intensity has an effect on feeding behavior in the damselfish, but aggression fluctuates little during the day. As the light intensity decreases in the afternoon, the fish stay closer to their shelter, and this in turn tends to raise the number of confrontations, challenges, and chases. This type of activity lasts until sundown, when retreat to shelter is in order. For many species, the shelter is not a territory to be defended, but rather it is any convenient place to hide from predators.

Some damselfish are territorial throughout the year, but reproductive activity intensifies this aggressiveness, since the territory becomes so much more important.

The damselfish is so aggressive in territorial defense that it will even drive off a larger and apparently better-equipped foe like a

crab. This behavior may be instinctive, or perhaps the defender has learned that the crab might prey on damselfish eggs.

Aggression is rarely displayed at such times as feeding, unless it is against a member of the same species. But none of the viciousness displayed in nest guarding, for example, is apparent when a predator enters the feeding area. The word "territory," however, has a different meaning, according to the intruder. The bicolor, for example, actually has a number of territories positioned as concentric rings. Certain nonthreatening species are permitted to enter the outer portions of the territory and are attacked only when they intrude into the inner regions. One intruder never challenged is the small harlequin bass. Egg-eating predators such as the wrasse, surgeonfish, parrotfish and others, however, are driven off upon entering the territory before they have the opportunity to approach the nest too closely.

Characteristically, fish guarding eggs will nip and drive off intruders, even those several times larger, and even ripe females, if the male is busy with nursery activity. An average nest contains about 350,000 pale, elliptical eggs with sticky tendrils at one end to help attach the eggs to the surface of the substrate. These nests are what attracts the predators. But the guarding males vigilantly watch the nests until just before hatching, up to a week after spawning. The male may remain near the nest after the hatching, but then usually abandons it within hours.

Territorial defense. A photographer, wanting to take a picture of a hermit crab in its snail shell, carefully installed the crab on a piece of coral. When he positioned himself and looked through the lens, the crab was gone. The photographer set the crab back on the coral and retreated once again. And again the crab was gone. This happened until the photographer got back to the camera fast enough to shoot this sequence showing a damselfish pushing and lifting the intruding crab off its piece of coral.

Social Status

The reef habitat of the bicolor—and most other damselfish—is unusual in that so many animals of different species live together in such close proximity. Thus the aggressive and defensive behavior can be divided into an attitude directed at outsiders and another one directed at members of the same species. Defense of eggs, for example, is usually directed at members of other species, while territoriality is usually aimed at conspecifics.

In the bicolor damselfish, for example, the intraspecific aggressive attacks upon members of the same species are closely linked with a well-defined social structure, marked by a hierarchy of individuals or dominance patterns and specific territoriality.

This strictly structured hierarchy is based primarily on size, and since the males of the species are larger than the females, it is based on sex also. But it is not a straight-line correlation between size and status, since smaller fish on occasion rank higher than larger fish. When the fish are busy establish-

Bicolor damselfish. *A highly regulated social structure can be found among the bicolors, usually with the biggest male as top-ranking individual. The bottom fish, usually the smallest female, serves as a whipping boy for the rest of the community.*

ing territories, the aggression, in terms of challenges and chases, is usually directed at individuals of nearly similar rank. Thus, the No. 1, or alpha, male is more likely to engage, or be engaged, in a chase or challenge with the No. 2, or beta, fish than he is by the lower-ranking individuals. The lowest-ranking fish, the smallest female, is referred to as the omega fish.

Countershading. *The brilliant coloration of many damselfish is not always constant. The colors may change for any number of reasons, such as to show increased aggressiveness or to display a readiness to partake in spawning.*

The pattern, which involves two types of confrontations, chases and challenges, is, of course, affected by the positioning of the fish, since the territories are not correlated to rank. Thus two "neighbors" may not necessarily be next to each other in rank. And, too, the alpha male roams freely about the area and thus has more opportunity to be confronted by a wider range of individuals than does the more sedentary subordinate. And even though the preoccupation happens with closely ranked individuals, observers noted that the alpha fish chases every individual, and every individual chases the omega fish. In turn, every fish except the little omega challenges the alpha, and no one challenges the omega fish. In other words, the tendency is to challenge those higher in rank; chase those lower.

The females wander throughout the neutral territory, but the lower in rank are often chased out of the defended areas by high-ranking females. There is no territoriality exhibited by females toward males, and the territories of the females are not very well defined because of their almost constant wandering.

This dominance pattern changes, however, once the territories have been established and reproductive activity begins. The females, for example, while still subordinate to the males, show an increased willingness to challenge each other, and even the males.

Among the males, during the reproductive period, there is a decrease in chasing activity as they are busy with courtship activity and guarding the eggs. The rankings also change sometimes. An observer noted that during the reproductive period a smaller fish replaced a larger one as the No. 2 male in the hierarchy. It was also during this period that the number of challenges by subordinates increased, indicating that the aggressive behavior induced by spawning activity overcomes the learned order of social structure within a colony or population.

This system of dominance has an unusual aspect to it: the omega individual seems to serve as a scapegoat for the entire community. This small fish is chased so much that it spends much of its time in hiding. The chasing in this case is not at all random, for higher-ranking individuals seem to seek out the omega fish for abuse, especially if they themselves have just engaged in some aggressive behavior and have been frustrated. The omega fish, usually a female, is not courted like the other females.

Observations of the bicolor damselfish social structure have been made in both laboratory tanks and in the ocean, and the behavior is basically the same under both conditions.

The omega fish, then, seems to serve the function of allowing those above it in status to release aggression and retain a more peaceful community on an overall basis.

Behavioral differences. Virtually all damselfish have social interaction with other species members, though without the rigid organization of the bicolors. There have been isolated individuals, however, who seem to have little contact with conspecifics.

Courtship

Behavior, whether instinctive or learned, can be precipitated by any number of different types of triggers called "releasers." Releasers may include color changes, speed of swimming, size of objects, or sounds. The releasers are as many as there are fish in the sea. The bicolor damselfish, for example, uses sound as a releaser to elicit courtship behavior. And this very social fish can be stimulated to courtship behavior by the playback of recorded mating sounds, showing that artificial releasers can elicit programmed behavioral acts.

The bicolor damsel, as its name indicates, is essentially a two-color fish, which usually has a distinct line of demarcation between the darker, bolder front portion and tail fin, and the lighter posterior. The front and fin colors are in the dark-blue, brown, or black range, while the back portion of the body is a pale grayish white to pallid yellow. Although the bicolor pattern is what helps distinguish *Eupomacentrus partitus* from other damselfish species, the bicolor does exhibit the greatest variation in color patterns in its family.

The color pattern is not always constant. During the first stages of courtship behavior, the darker areas deepen slightly. When the courtship activity intensifies, these areas are extremely dark and are actually reduced in size, limited to just the head and snout, and the tip of the tail. This is called the "black mask" pattern. When the behavior is aggressive, rather than courtship, however, the dark areas increase, covering the whole body except for a narrow pale stripe on the flanks. This is called the "white bar" pattern. Since this color pattern change is so distinctive during courtship, it is easily tested and observed, as is a certain swimming movement which is actually associated with courtship, the "dip," which is an obvious down-and-up swimming pattern.

The bicolor damselfish, especially the territorial male, produces an array of sounds, which include a chirp, a long chirp, a grunt, a burr, a poplike sound, and stridulation. Some of these are made during courtship activity while others are associated with feeding or aggression.

When the sounds were recorded and then played back to fish in a laboratory, the females showed no change from normal behavior. But when the males heard the chirp and long chirp sounds, there was an obvious increase in their dip movements and an in-

crease in producing their own chirps and other sounds usually associated with courtship. This courtship behavior was induced even when reproductive activity would normally be at a minimum. It was noted, interestingly, that even though this behavior was induced by an underwater sound projector, no fish directed any of its activity toward the sound source, always to another fish.

As with the dip movements and sonic responses, color changes were induced by the recorded sounds, first the slight deepening of the darker colors, as in low-intensity courtship activity, and then the "black mask" pattern.

When tests were conducted at sea, the tape-recorded sounds were once again played back, and once again the various chirps elicited an increase in motor, sonic, and chromatic responses associated with courtship behavior. This experiment with recorded sound also showed that the playing of a "pop" sound, usually indicative of aggression, inhibited courtship activity.

Clownfish's Living Home

A very special interrelationship exists between clownfish and anemones. The clownfish belongs to a branch of the damselfish family which is found mainly in the Indian Ocean and western Pacific. Many of the clownfish species that inhabit anemones are what scientists call "obligate symbionts," which means they probably could not live an entire life without the anemone.

Anemones usually prey on small or dead fish, but various species of *Amphiprion*, or clownfish, are able to live among the anemone tentacles even though the tentacles are covered with venomous stinging cells. Considerable variation exists in the relation-

Anemonefish. *Several species of* Amphiprion, *a variety of damselfish, live among the tentacles of sea anemones without showing any ill effect from the anemone's stinging cells.*

ship, with some fish living with only one species of anemone while others show no specific preference. Some of the clownfish are very territorial, driving off intruders from their personal anemone. Eggs are laid at the base of the anemone, and the parent often rubs the anemone, causing the tentacles to extend, which in turn forms a protective canopy for the eggs of the fish. It is also common for many clownfish to feed their host, but they sometimes do just the opposite, sticking their heads into the mouth of the anemone and stealing food. The *Amphiprion* may even carry dead fish to the anemone. Some clownfish have been observed feeding on the anemone's fecal material and even on its tentacles.

Helping with the food. Some anemonefish have been observed bringing food to their anemones, or driving other fish into their tentacles, thus providing a meal for two.

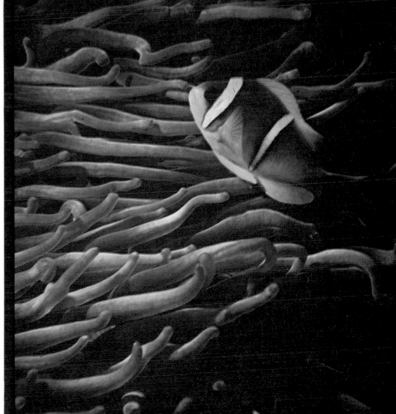

Special Problems and Situations

When danger threatens or man interferes with normal routines, the damselfish answer with a combination of instinctive and learned reactions. Most fish are constantly on the alert for predators, but have special responses. The blue damselfish, for example, shows little desire to engage predators and usually flees as soon as it is alarmed, heading straight for the closest shelter. This escape response is so stereotyped that it may even serve as a sort of warning signal for other species of reef inhabitants.

Mackerels, snappers, and jacks are the kind of fish that elicit this response, which occurs when the predators are still four to eight yards away. Such fish as the trumpetfish and lizardfish, which prey from ambush, are quite successful in their attacks on chromids. The escape response of the blue damselfish proves inadequate to deal with ambush predation.

While damselfish are always alert for predators, there are a large number of fish in the area which do not trigger the escape response. The blue damselfish is somehow able to distinguish between the dangerous predators and other fish. In undersea observation swimming speed, generating rhythmic pressure waves, seems to justify this distinction, but in aquarium tests size and color of models had more importance in precipitating the escape behavior.

When ordinarily nonpredatory fish do provoke the escape response, it is usually because of a radical change in behavior, such as

Out of the ordinary. Damselfish have developed a number of responses to threats and unusual situations. Whether these responses are totally instinctive, or whether some have been learned has not yet been fully determined. This is one example of the remarkable variety of ways in which these fish interact with their environment.

the emergence of a fish swimming rapidly toward them or of some other action which may be misconstrued as an attack.

The escape response is thus very complex. However, the exact relationship between an animal's instinctive reaction to stimuli and its learned behavior about predators still has to be determined.

Some anemonefish have been artificially deprived of their anemones, and in this totally unpredictable situation they proved they have certain innate behavior patterns. They often dig holes and feed them, defend them, even sleep in them, much as they would behave with an anemone. Instead of digging holes, some clownfish without anemones in an aquarium will settle into algae and direct behavior toward it similar to behavior directed toward anemones. Other clownfish have been observed bathing in bubbles from an air stone and defending the bubbles, probably seeking stimulation comparable to that of the tentacles of the anemone.

Heading for Home

Damselfish display a strong instinctive urge for their homes once they have chosen a particular piece of coral or rock or, as in the case of clownfish, an anemone. If a clownfish is separated from its anemone for a while and then reintroduced, it must go through a painful period of reacclimatization. When the fish again makes contact with the anemone, usually with a fin, the anemone's venomous nematocysts are discharged and the fish jerks back the fin. The anemone's tentacle then sticks strongly to the fish, and probably hurts it, but gradually the contacts by the fish are increased in both frequency and area exposed. Somehow the little clown becomes less vulnerable to the poison and is soon wallowing among the tentacles. It is now known that the clownfish's immunity is not developed because the anemone finally adopts or recognizes its tenant, but rather by a change in the mucous covering of the fish, which inhibits the discharge of nematocysts. If deprived of their anemones, some clownfish may seek shelter in coral. Others, when they are in an aquarium, isolated from their anem-

one, will wallow in the rising air bubbles as a substitute for the anemone itself. Other damselfish, such as some species of *Dascyllus,* live among anemones as juveniles but live in coral as adults. This close association with coral is very strong among many species. Some juvenile *Dascyllus* display no special attraction for certain types of coral. But once they make a choice, they show a preference for that species from then on. Thus it is through early experience that the fish develops a preference that will affect its behavior for the rest of its life.

There are times when an instinct, such as the need for shelter, and learned behavior, such as a preference for certain coral, combine to put an individual in a difficult situation. For example, damselfish hide among coral and remain in it when a piece is broken off and taken from the water by a diver. This is a clear-cut case of the instinct for self-protection overriding everything else.

Home *is where the coral is. Some species of the damselfish family learn exactly which coral is their own and if relocated will choose another home in the same species of coral.*

Chapter VI. Change the World or Change Yourself

Modern man can alter some aspects of his environment to suit his needs. He dams rivers, chops down trees, paves vast areas, burns grasslands and forests. Other forms of life, however, (and even man himself until recently) are unable to adjust the environment. Instead they must adjust their behavior, or even their forms and functions. This is especially true in the sea, where conditions of temperature, light, pressure, and many others are harsh and changeable.

"The sequence of days and nights requires adaptation by many fish species. Some adjust their activities, their colors, their whole way of life from day to night."

When the tide ebbs, many creatures are left high and dry. If they are unable to meet these drastic changes, they will perish from overheating or dry out under the sun. Two such animals of the tidal zone are the blue mussel and the periwinkle, both very common along the Atlantic coasts of Europe and North America. Both molluscs have adapted in their behavior and in their forms. They can shut themselves within their shells to prevent dehydration and thus survive many hours in air until the tide covers them again. To keep from being battered in the surge of moving tides, the periwinkle ensconces itself among rocks and the mussel secretes threads of a sinewy material called byssus that binds it to its substrate.

Another drastic behavioral adaptation is exemplified by sea stars, which are free-swimming when they hatch. Upon reaching adulthood they settle to a slow-moving benthic existence. Some marine fish live around estuaries where the flow of fresh and salt water alternates with the tides. These fish must be able to adjust their osmoregulatory systems so alterations in the salinity of the water will not upset their internal chemical balances. Seasonal changes in the environment send some creatures to more hospitable waters or into hibernation. The cunner, a wrasse of northern temperate waters, disappears each winter. Whether it goes to deeper water with more consistent temperatures or hibernates in the mud is unknown, but it uses one of these methods to cope with changes.

The sequence of days and nights requires adaptation by many fish species. Some adjust their activities, their colors, their whole way of life from day to night.

More and more, man's adaptations of the environment to his own desires have played cruel tricks on other life-forms. When man began dumping heated water from his power plants into the sea, it warmed the waters and caused luxuriant algal growths that could choke out other life. The heated water also attracted animal species that needed warmer waters and which would normally have migrated in winter. When the plants close periodically, the waters cool and fish adapted for warmer waters die.

To continue to exist on earth, life must adapt to new conditions or die. We have not yet fully realized how much man's ability to change the environment has increased our responsibility to the entire living community—and to our descendants.

Emperor penguins. A parent bird watches over a downy chick, ready to protect it from antarctic winds reaching up to 80 miles an hour and temperatures as low as 80° below zero. Penguins often huddle in a V formation to help break the wind and to share body heat.

The Influence of Light

Light is one of several stimuli that greatly affect the behavior of many creatures, marine animals in particular. Some are attracted to it, others repelled by it. (See the volume titled *Window in the Sea*.)

Many animals use light as a means of orienting themselves to their environment. This is called phototaxis. Light also informs an animal of the time of day or its location in relation to a reef or cave.

Some fish demonstrate their phototropic behavior by swimming toward light. Others may simply present their dorsal sides toward the light source. Since sunlight comes from above, fish usually swim with their dorsal sides up. But in a submerged cave with a sandy bottom, sunlight may reflect off the white sand bottom and provide more light from below. As a consequence, some fish turn over and swim upside down. Or if light comes from the side, a fish may lay over sideways to present its dorsal surface to the light.

Fish sense light with their eyes. Some species also utilize a "third eye," or pineal organ, atop the head. Some scientists believe the pineal organ was once a sort of eye and, in fact, it can sense light even with a growth of skin over it. Hagfish, some sharks, and halfbeaks are among those with a third eye (pineal organ). These organs lie within the skull, beneath an opening in the bony brain case and a layer of unpigmented skin. Hatchetfish, which are repelled by light (photophobic), sense the lower light levels above them at night and ascend into surface waters to feed. By day, they remain deep in the sea where little light reaches them.

Squirrelfish swim upside down, orienting themselves to the source of light which in this case is a reflection from the white sand bottom of the cave.

Building a Life Raft

Most snails trudge along a solid substrate on the sea floor, foraging as they move at a snail's pace. *Ianthina,* the violet snail, however, has adapted to a different way of life that makes available to it food that other snails cannot reach. *Ianthina* builds itself a raft of mucus-coated bubbles. On its raft, it drifts before the wind and on ocean currents much the same as its favorite food, *Vellela,* the by-the-wind sailor, and *Physalia,* the Portuguese man-of-war.

Ianthina is born alive, a tiny snail 1/300th of an inch in diameter and with a thin shell that is buoyant. As the animal grows within the shell, it becomes heavier than water, like other snails. To remain afloat, *Ianthina* has developed the ability to trap air in the cupped end of its foot and secrete mucus to form bubbles. It presses these bubbles together forming a raft that provides the buoyancy it needs to live its unusual floating snail's life. In fact, if this snail is separated from its raft, it will sink and die, as it would be unable to obtain air to build another bubble raft. *Ianthina*'s ability to move by itself and to feed as other snails do on the substrate has been reduced to near zero.

As it drifts, *Ianthina* hangs suspended from the raft with its two-forked feelers and proboscis directed toward the surface where food will be found. When the tentacles contact food—*Velella, Physalia,* or anything else edible—the sightless sea snail bites into the flesh with its proboscis and eats as it is carried along on the sea's surface with its catch. *Ianthina* not only seems immune to *Physalia*'s powerful poisons, but seems to mimic it by emitting streams of purple dye.

Purple sea snail. *A self-made life raft constructed of mucus-covered bubbles enables the purple sea snail to drift through the ocean in search of food.*

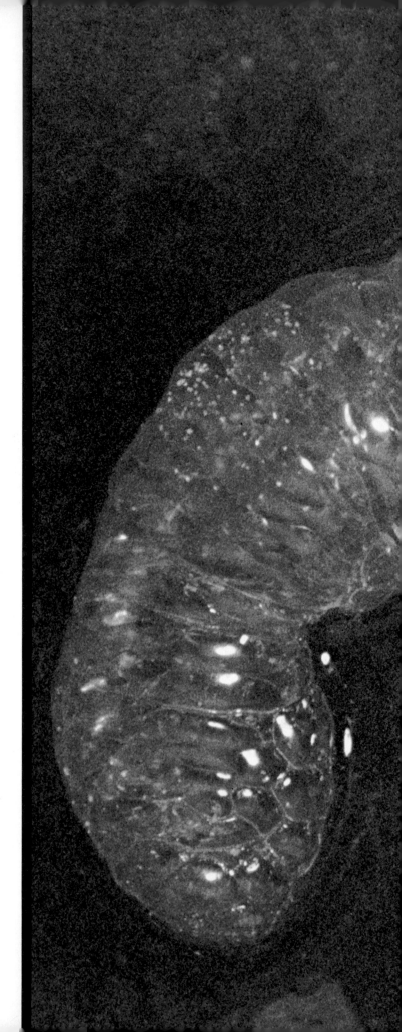

Housing Shortage

A fish in the open sea has no place to hide. It must depend on its coloration to become less visible, on its senses to be aware of the approach of predators, and on its swiftness to escape when it detects them. This lack of hiding places may be why fish commonly aggregate beneath floating objects at sea. The shadow under flotsam can make a fish less visible, and the fish seem to know it. In one study, darker-colored fish were found close beneath floating debris where shadows are more intense. Their distribution suggests that fish can pick the best hiding places for themselves.

Flotsam may also attract fish because it offers a point to which they can orient themselves. Thor Heyerdahl, sailing across the Pacific Ocean on *Kon Tiki,* reported a popu-

Floating home. *Above, these juvenile fish are attracted to a leaf. It probably gives them a feeling of protection. Later, as the fish mature, they will abandon such primitive means of support.*

Hiding place. *Right, the branches of a tree provide cover for several triggerfish. The large size of the tree and the complex pattern of its branches and their shadows enhance the tree's attraction as shelter.*

lation of fish beneath the balsa raft. They ranged from the smallest juvenile forms to large predators like tunas, dolphins, and marlins. Once *Kon Tiki* even attracted an enormous whale shark. Some fish took up residence between the raft's logs where they could be safe from the predators.

The list of floating objects that are found at sea is a long one, ranging from logs, coconuts, and trees washed out to sea from rivers, to sides of tumbledown buildings, shipping

crates, and parts of boats. In some tropical seas, huge patches of floating seaweed provide cover for many animals, some living in the seaweed and others swimming beneath it. Fish commonly gather beneath rafts provided for swimmers, and even anchored boats have populations of fish under them. This tendency of fish to associate with floating objects has been used to build new marine communities. Moored objects deliberately set out attracted small fish first, then increasingly larger fish joined to feed on the early arrivals. Even wide-ranging species paused for a time under the cover.

Many fish live in close proximity to much larger fish. Sharks have pilotfish and remoras with them. Yellowfin and skipjack tuna commonly school close beneath porpoises off the California coast, but tuna fishermen have learned of this behavior and seek the porpoises as a telltale of the tuna.

Life beneath the floating debris in the open ocean appears to be a defensive behavior for some fish and a feeding behavior for others.

91

Harsh Living

Living creatures go through drastic behavioral changes to survive. These adaptations often characterize a species.

Mussels, for example, attach themselves to rocks and other solid objects by means of strong threads, called byssus. These threads are secreted internally by the mussels as a sticky substance which hardens and becomes extremely strong after coming in contact with seawater. These threads help keep the mussels from being battered against the rocks during times of heavy surge. And because they are so sessile, mussels mate by ejecting their eggs and sperm directly into the water, where fertilization takes place.

Since mussels live in a tidal environment, they must adjust to the ebb and flow. Experiments show that they possess an internal rhythm that is correlated to the tides. Variations in the pumping of water through the gills for feeding coincide with the tidal variations. This correlated rhythmic behavior helps conserve the organism's energy. This rhythmic pattern may be dictated by some internal time clock, for even when mussels are removed from a tidal environment and placed in an aquarium, the rhythmicity of behavior continues.

Another type of adaptation for survival is the homing behavior of the limpet and false limpet, two univalve molluscs which live on rocks in the tidal zone. These snaillike animals, resembling miniature volcanoes, appear to be totally sessile, but when the tide comes in and covers them, they wander from the home base, a scar they have excavated on the rocks they live on, to graze. One species travels only about six inches from its home scar in search of food, while another ranges as far as six feet. As the tide ebbs, these creatures find their ways back to their places on the rocks.

In one experiment, to determine how limpets and false limpets navigate, the area between a wandering animal and its home scar was wire brushed and rock was chiseled away. With these chemical and topographic features altered, the animal was unable to find its way, indicating limpets use these signs for homing to their assured niches on the shoreside rocks.

Limpet. At left, the circular home scar of an Acmaea limpet is bared as the half-inch mollusc starts its inches-long journey over the face of the rock, grazing on algae as it moves.

Blue mussels, right, possess an internal rhythm that is correlated with the tides. Changes in feeding behavior are linked to tidal variations and are rhythmically performed even when the mussels are placed in the unchanging environment of an aquarium.

Sleeping Fish

As darkness envelopes the coral reef, the night shift of animals emerges and daytime creatures become subdued. They settle in for a night of inactivity, conserving energy as they rest. Some enter into a state comparable to man's sleep. Other day animals simply lie quietly awake and alert but otherwise shut down for the night. Angelfish, for example, lie quietly in narrow crevices, protected from predation although less aware and less able to see in the darkened waters. Some wrasses bury themselves in the sand. Some parrotfish lie on their side, fast asleep. Other smaller species of parrotfish nestle into niches in the coral rock and secrete protective mucous envelopes about themselves. These webs are so delicate that the slightest disturbance in the nearby waters ruffles the mucus and alerts the sleeping parrotfish.

*A Pacific species of **surgeonfish,** member of the tang family, lies quietly against a colony of feeding coral polyps, some of which cover the fish's dorsal fin.*

Among the members of the tang family, the surgeonfish is an active daytime feeder, traveling in large schools close to the bottom in reef areas. By day the schools of tangs are alert to any danger that approaches. At night these fish abandon their algae grazing and their large schools; they disperse and go to rest, well hidden in their home reefs. Being laterally compressed, tangs often find narrow crevices much as angelfish and butterfly fish, and lie there protected through the night. Some nestle up to coral heads and among the swaying branches of alcyonarian corals for protection, becoming less visible while they rest. Their retractable peduncular spines continue to serve the fish as effective defensive weapons.

These diurnal changes of behavior trigger physical changes that alter the appearance of both day and night feeders. The daytime population that lies quietly through the night, or moves in much more subdued fashion, take on less brilliant coloration. Stripes and blotches become fashionable as a key to becoming less visible. Butterflyfish, for example, take on vertical stripes along their sides which break up their outline and help them blend with the staghorn coral they frequent. Quick-changing hogfish adopt a somber, blotchy hue to match their backgrounds.

The nighttime feeders too change from daytime behavior patterns and colors to those that better enable them to hunt their prey undetected. Spiny lobsters, which hide in narrow overhangs by day with only the tips of their antennae showing, emerge to march openly about the reefs in search of other crustaceans and molluscs to prey on. Basket stars, which are bunched up like so many fists by day, expand into delicate lacy creatures by night that strain planktonic food from passing waters. Squirrelfish, bigeyes, and cardinalfish leave their daytime lairs and actively seek food by night, gathering dim light in their large eyes far better than the smaller-eyed daytime population. Octopods and squid, which can change color and pattern faster than any other living creature, move into action. Bright moonlight also has an effect on marine creatures, causing some that normally rest at night to keep active. Triggerfish that feed over bright sandy areas continue foraging in moonlight, while other triggerfish that find food on dark rocky surfaces do not. They retreat to their small holes, lock their dorsal fin spines in place, and rest until daylight.

*This surgeonfish, another species of the **tang family**, swims actively about during the day, feeding. Tangs often move about in large schools during the day.*

Color Changes

Goatfish change color and pattern dramatically as a defensive adaptation. Whether by day or night the various species of goatfish blend themselves so well with their backgrounds they defy detection. Because they are so easily overlooked by predators they are able to survive as they grub around for their food. Typically, goatfish use their paired barbels, which extend downward from their chins, to taste in bottom sedi-

Yellow goatfish. They seldom cruise through midwater, as they are bottom feeders; but small groups like these goatfish have been observed traveling fairly long distances and are probably still young.

ments for the invertebrate morsels they eat. Their barbels work simultaneously as tongues and as walking fingers which twiddle back and forth to stir up food. Because the activity of feeding would surely detract from their cryptic coloration, they must remain alert while searching for food. Color changes do take place, though.

In particular, the spotted goatfish adapts by assuming a mottled pattern of dark and light patches that break up its fishlike outline and make it harder for predators to recognize it. Similarly, spotted goatfish can adopt an almost white appearance to match the coral sand they often stir up in search of food.

Young spotted goatfish have silvery sides and dark backs, a color scheme that serves them well because as juveniles they cruise the midwater regions. Only when they reach a certain size range and find a proper sort of sea floor do they assume their adult color, pattern, and behavior. They then descend from midwater and reside along the bottom.

The goatfish's ability to change color and pattern quickly and drastically as much as its exquisite taste made it a favorite at the sybaritic Romans' dinner tables. Wealthy gourmand Romans often kept goatfish in specially built salt ponds so their dinner guests could choose their live fish and watch them undergo color changes as they died. Except for the popularity it gave them with the Romans, their color-changing ability serves goatfish well.

Goatfish have made other adaptations that have enabled it to survive. Being a bottom feeder, its mouth is ventrally located. As it cruises along the bottom, feeling and tasting with its barbels, its mouth is in a position to suck up any animals sensed moments before with the barbels.

Many other families of fish have the ability to change color. It is not uncommon that brightly colored diurnal fish change to drab colors at night, and that nocturnal fish that emerge when light wanes adopt color patterns best suited to the circumstances.

Goatfish. Outfitted for nighttime feeding on the coral reefs, a spotted goatfish dons red and white mottled pattern as it rests on the bottom during the night.

Fur seal. Abundant food and a safe place to have its pup brought this northern fur seal cow to St. George Island in the Bering Sea.

Migrations

When the weather turns cold and the days short, many animals move to warmer climes where procreation will be accomplished safely. When warm weather returns to the temperate zones of the earth, these same creatures move back. In the Northern Hemi-sphere, the migrants move north in spring and south in fall. These movements are adaptations the animals have made, enabling them to be in the most advantageous places for their survival and perpetuation of their species.

Northern fur seals spend much of the year scattered over the Pacific Ocean, but as

Heermann's gull. *While most birds migrate north in spring, these Heermann's gulls move south for breeding in the Gulf of California.*

spring approaches they gather in large herds to move north to their island breeding grounds between Alaska and Siberia. How and why these seals travel 3000 miles or more can probably be accounted for by race memory—knowledge that has led the animals to salutary feeding and breeding grounds for many generations.

Two species of birds sometimes migrate opposite to the expected direction in pursuit of favorable climates. Some Heermann's gulls winter along the west coast of the U.S. and Canada and some elegant terns winter in California. In spring they migrate south to their traditional breeding grounds on Isla Raza in Mexico's Gulf of California.

Inhospitable Shores Teeming with Life

Sandy beaches along seacoasts and estuaries seem most inhospitable. Despite this, the habitat supports a considerable population of animals. Depending on the state and activity of the tide, the stresses of the beach environment include the washing away of substrates and deposition of sediments. Aquatic animals living on intertidal beaches face the possibilities of drying or drowning and of osmotic imbalances as fresh water and seawater come and go with the tides. They may be preyed upon by terrestrial as well as aquatic animals, and they may be subject to poisoning from lack of oxygen or high levels of ammonia or hydrogen sulfide from natural or man-made pollution.

Other shoreside environments that house many organisms are mangrove beaches and swamps. These areas are not as inhospitable as barren sandy beaches; in fact, they teem with life. The large population of small creatures attracts hosts of larger predators.

Fiddler crabs often gather in large hordes on sandy beaches that are punctuated with growths from mangrove roots called aerophores, small shoots that pop up through the sand into the air. The crabs gather in these areas because of the wealth of food but they must avoid being devoured by larger, hungry citizens of the mangrove jungles. The washing away and depositing of sediments occurs on a reduced scale because of the holding qualities of the mangrove aerophores, but the sedentary life of the fiddler crabs and others is still disturbed.

To survive in such environments the shore crabs have had to make many behavioral adaptations including burrowing and burying to protect themselves from dehydration, overheating, and abrasion by sand and rocks in the surf.

Another behavioral adaptation is directional orientation. Fiddler crabs have an uncanny ability to find their burrows after foraging on featureless beaches. Other adaptations help them anticipate tides and time of day so that they know when it is safe to be out of their burrows searching for food.

Chapter VII. Communications in the Sea

No man is an island, and no animal is either. All animals communicate to a certain extent, mainly to other members of the same species, but also to others.

By definition, any system of communication demands at least two individuals, a sender and a receiver. But this is somewhat limiting. There are many signals which are not directed toward a particular receiver but are, in a broad sense, communication. When the starfish is detected chemically by sand dollars and they bury themselves, it has indirectly communicated its presence. The receiving end of a natural communication system involves senses that were discussed in two earlier books: vision, hearing, lateral line, taste, smell, touch, and sensitivity to electric fields. The simpler the animal the more "automatic" the communication, triggered by stimuli we can easily identify. With animals whose behavior is influenced by learning, our explanations become unsatisfactory. And when we reach the level of toothed whales, which have a highly developed cerebrum, the sophistication of the exchanges defies our understanding.

Most fish may be able to tell a predator that it is ready to defend itself, to the death if necessary; that same animal might want to tell one of its own kind that it is trespassing on private territory and must get off fast. The signals in these cases may be the same, but are interpreted differently by the predator and the territorial intruder.

What might be considered aggressive signals from one male crab to another could be interpreted as a willingness to mate if they were sent between a male and a female.

The ability to transmit, receive, and understand signals may be innate or may be learned. But every animal can communicate, however simple or complex the message.

Communicative mechanisms might have developed from other behavioral patterns called *expressive movements* and may have been enhanced by the evolution of a prominent feature, such as an enlarged dorsal fin which is displayed in signaling.

"The ability to transmit, receive, and understand signals may be innate or may be learned. But every animal has this ability, however undeveloped or sophisticated."

There is a distinction between actions that may be communicative, such as shivering when you are cold, and those that are both communicative and expressive, such as those that tell another individual that you are ready and willing to mate. The shivering is an instinctive movement which could be developed into a differentiated expressive movement in special instances. For example, a child may shiver with cold and its mother may warm and comfort it. When the child finds that all it has to do is shiver—even if it is not cold—and the mother will come over and comfort it, the shivering has become a differentiated expressive movement.

Expressive movements are never totally the same, even in closely related species. In animal communication the signals must be obvious, yet precise; and they must be conspicuous without being easily confused.

American lobster. *This favorite of American diners is a very defensive animal and will fight most intruders on its territory unless it happens to be a female during the breeding season who excretes chemicals into the water, signaling her receptivity to a potential mate.*

Seeing-Eye Goby

A variety of pistol shrimp lives in a burrow and is very nearly blind, while its roommate, a fish of the goby family, serves as the shrimp's senses, or at least as a kind of seeing-eye fish. The goby lives in the shrimp's burrow, using it as a temporary shelter during the day and as a place to rest at night. The shrimp also cleans the fish. In return, the goby emerges from the burrow and acts as a guide and sentry for the shrimp, heading back to safety at the first sign of danger.

The two signal each other with a complex communications system, which includes the use of the shrimp's antennae and its pinchers, and the fish's use of tail fanning. The shrimp emerges from the burrow only in the presence of the fish, and the sequence, as observed in the sea and in an aquarium, involves the shrimp moving headfirst toward the entrance of the burrow with antennae raised until contact is made with the goby. This antennal contact is virtually constant throughout the emergence, while on the outside of the burrow, and during reentry. The fish acknowledges the antennal contact by gently fanning its tail. The shrimp then follows the fish, or if the goby is already outside, the shrimp then emerges from the burrow. If the goby is still inside the burrow and seems not to respond to the shrimp's antennae, the shrimp pushes the fish with its pincers or signals it by clicking its pincers.

During laboratory testing, the shrimp never emerges from the burrow unless the goby is at or near the entrance. If the fish is removed, the shrimp might be observed near the burrow entrance with its antennae raised forward, as though feeling for the missing goby. The shrimp fails to emerge after the fish has been removed, but it is finally induced to leave the burrow after being deprived of food for 48 hours.

When the shrimp leaves the burrow in the presence of a goby, it might stay as long as two minutes if it feeds or as little as several seconds if it only shovels debris outside the burrow. In retreating to the burrow, the shrimp always returns tailfirst. The fish also reenters tailfirst, except in times of great haste when it reenters headfirst.

In the absence of fish, when the shrimp has been lured from the burrow, visual stimulation has no effect on the shrimp; it is apparently unable to detect a threat by sight.

There are many other types of shrimp which use the antennae extensively in signaling. This is especially true of the cleaner shrimp of the genera *Stenopus* and *Periclimenes* which wave their long antennae to attract the attention of "customers" so they can provide a cleaning service similar to the one the pistol shrimp provides its goby.

Goby. *Top of page, a goby vacates the burrow while a shrimp performs the housecleaning tasks for the both of them.*

Guide. *Opposite page, top left, the goby emerges from the burrow preceding the shrimp. Mid left, the shrimp establishes contact with its antennae, then at bottom left responds to tail waving and emerges. Outside, top right, antennal contact is maintained at all times, even through reentry mid and bottom right, with the shrimp always backing in first.*

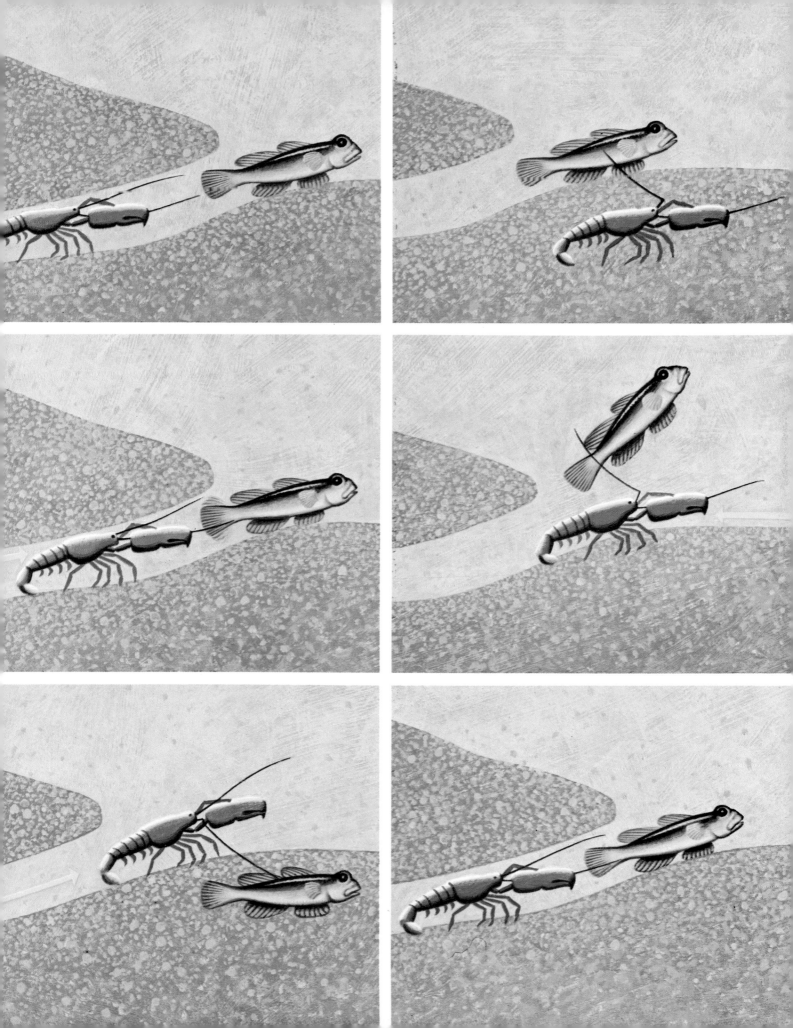

Specific Message

Communication may be very subtle, with the message intended for only a few specific individuals. This may be the case with the solitary coral reef fish which are often called "poster-colored" or "flag-colored" because of their gaudy coloration and bold patterns. Their spectacular coloration is not used for camouflage, since they can be spotted for quite some distance. In these fish, color serves some other purpose; it may be an isolating mechanism or a warning signal as it is in many types of poisonous fish.

Experiments by Konrad Lorenz, the noted behaviorist, led him to conclude that the bright coloration of many fish serves to make them very obvious to other members of the species, telling them, "Here I am, this is my territory." If the trespasser persists in staying around, a fight ensues. Lorenz said that in his aquarium it was almost always a fight to the death, but in the sea the intruder usually had the opportunity to escape.

In the small aquarium and in limited areas of the ocean, several different species of these poster-colored fish are able to live in close proximity, but generally with only one member of each species permitted. As soon as the territory holder spots a like-colored fish, it pursues and attacks.

We must be careful not to generalize too widely about the territorial aspects of bright coloration. In many other cases, such colors may serve as a warning that the displayers are poisonous, venomous, or simply distasteful, as are many beautiful nudibranchs. In a few cases, a species may mimic the bright colors of another as a means of defense.

Angelfish. The bright coloration of the queen angelfish, top, and the French angelfish may serve as a warning to territorial intruders of the same species.

Open-Water Cleaners

In the delicate relationship between cleaner animals and client, or host, fish, the need for communication is fairly obvious. The fish that wants to be cleaned has to make it known, and this may be done with a certain body posture or, like the goatfish, with a change in coloration. Or the host may signal only a certain part of the body to be cleaned by presenting that part of the body and making it accessible for cleaning.

Most of the "cleaning station" activity has been observed in the warmer waters around coral reefs where small, colorful fish and shrimp pick parasites off larger client fish. There have been observations among some fish, butterfly fish for example, of members of the same species performing cleaning operations for each other. Generally the smaller, specialized animal performs this function for a larger host fish.

The fish that wants to be cleaned will swim up to a "cleaning station" and stop, fan its fins, perhaps float at an odd angle, and even change its color, thus communicating its desire to be cleaned. The cleaners then proceed to climb or swim all about the host's surface, picking and eating small organisms such as copepods which they might find among the scales, fins, gills, and elsewhere. Sometimes the client will open its mouth and the cleaner will proceed inside without hesitation to do the work required there.

A typical cleaner-client relationship exists between the butterflyfish and the goatfish. If a goatfish properly signals its intentions by lowering its head and changing its body coloration, both of which the butterflyfish interprets as nonaggressive signals, the cleaning will take place and continue as long as the signals stay in effect or until the larger fish swims away.

This cleaning symbiosis, however, has rarely been observed among offshore or open-water fish. But such observations were made off Point Conception and in Monterey Bay, California. The giant sunfish *Mola mola* had been following a large number of lion's-mane jellyfish in a cold-water layer beneath the surface. Divers observed several sunfish stopping, vigorously fanning their fins, and tilting backward at about a 30° angle. The fish were not at all alert and were in an ap-

parent trancelike state. One observer said they were "mesmerized." But soon the sunfish were surrounded by perch, who began pecking at their eyes, mouths, nostrils, gills, and fin coverings, picking off tiny red worm-like organisms. After the cleaning operation, the sunfish swam off, showing no apparent effect from the "trance."

This behavior was noted more than once, where "stunned" or "mesmerized" sunfish were watched in the colder water and sig-

Cleaning signal. Effective communication is needed when a fish wants to be cleaned. These goatfish change color and lower their heads to indicate their readiness to be cleaned by a butterflyfish.

naled their desire to be cleaned by fluttering their fins and tilting backward at an odd angle. Immediately the perch appeared and proceeded to clean up the red organisms. And once again, the sunfish seemed to come out of their stupor and were able to swim away into deeper water.

Meeting of Stars

Two sea stars, *Patiria miniata,* meet in the coastal waters off California and raise their arms as if in a greeting. What are they telling each other? Only the sea stars know. But certainly this reaction suggests that some communication exists between them.

Classified as "lower animals"—animals without backbones—sea stars are also without heads. Instead they have sensory organs on the tips of their arms. Nerves radiate from the body to the arms to operate the animal's tube feet and sensitive feelers. The feelers are able to receive vibrations through the water and are particularly sharp at distinguishing chemicals dissolved in it.

The taste preferences of various members of the 2000 or more species of sea stars are becoming increasingly well known as researchers observe these slow-moving creatures. Some subsist on a diet of molluscs, others on shrimp or fish; one specializes in eating coral polyps, and still others eat their own

kind. The sea star's sensing organs are astute enough to tell them that their favorite food is near. Some mollusc eaters can tell that there is a broken oyster in the vicinity and move in its direction without hesitation.

Like other marine creatures that release their sexual products into the water, sea stars must do so when another of the opposite sex is nearby and ready to spawn. To coordinate their reproductive activities sea stars need to be able to communicate, otherwise the eggs or sperm will be wasted.

It has been established that many species of sea stars have an individual chemical make-up by which they can be identified. Prey of one species of star will try to escape from only that specific star. This leads us to believe that sea stars are able to recognize and communicate with each other chemically.

The tips of sea star arms are equipped with light-sensitive organs. When two stars meet and curl up an arm, they could be telling each other, "I see you, I recognize you, I will not eat you, I am ready to mate with you."

Chapter VIII. Inborn Skills

Many creatures come into the world programmed to meet the challenge of life. Inborn responses to the events of everyday life are called instincts, the "mechanical" unthinking reactions to the direction of light, the time of day, the threat of predators, temperatures of air and water, and other external stimuli. By having inherited behaviors, an animal needs no lengthy instruction period from a parent or trial-and-error learning in order to survive. The creature will automatically carry out the function that goes with a particular stimulus at the appropriate time, directed by inner drives.

"Creatures flexible enough to adapt quickly are far more likely to survive than those depending solely on instinct."

Instincts are passed on to offspring through the genetic material of the parents who received them in the same manner from their parents when they were conceived. While instinct offers a shortcut to learning that is beneficial in many respects, it has drawbacks that can spell doom for a species under certain conditions. The green turtle migrates from the coast of Brazil 140 miles to Ascension Island each year to reproduce. If these migrating turtles represented the whole population, they would be doomed if Ascension were to sink below the surface or become an active and uninhabitable volcano. The turtles' instinct would direct them to the middle of the Atlantic where an island used to be. There would be no reproduction and the species would be extinct within a few years.

To learn, an animal must acquire knowledge or skill and react according to a new program, not an inherited one. Learning ability is a necessary but insufficient factor for the development of intelligence. But if the animal can associate cause and effect, it will then act as circumstances demand.

Those creatures with enough flexibility to adapt quickly are far more likely to survive than those that depend solely on instinct, assuming they learn before they fall victim to predators or to the elements. Both intelligence and instinct have their advantages—neither way is the best in all cases. One way we can differentiate between instinctive and learned behavior is to raise an animal in isolation, protecting it from others that could teach it and from the means of learning by trial and error. If under these circumstances an animal responds appropriately to a model of a predator, mate, food, or other stimulus, we can conclude that its response is instinctive. Such a stimulus is called a releaser. The animal's reaction is one programmed into it at conception and will manifest itself only when given the proper signal. In addition, there must be a built-in readiness to act or react. The proper food at the wrong time or the competitive male at the wrong season will not elicit a response because there is no inner readiness to react.

Each species has its own set of signals that only others of its species are programmed to respond to. It is this mechanism that helps prevent breeding among different species. Instinct helps a species survive under normal conditions.

Sea snake. Pacific Ocean predators instinctively avoid venomous sea snakes, even when their appearances are altered. Atlantic predators, unfamiliar with these Indo-Pacific reptiles, attack them and often die from such encounters.

Sea anemone and hermit crab. Crab, above, touches the anemone, inviting it to move to the crab's new shell.

Response. Anemone responds in the photo below by taking hold of the crab's shell with its tentacles and raising its pedal disc.

Mutually Helpful

The Mediterranean hermit crab and a particular type of sea anemone have an instinctive symbiotic relationship even though each is able to live normally without the other. In most mutualistic relationships, the two organisms need each other for some important aspect of their existence such as procurement of food, attraction of mates, or protection from predators. In this case, the relationship is related to both food and sur-vival. The sea anemone becomes more mobile, increasing its chance of food capture, and the crab is protected from octopods. Hermit crabs are a favorite food of octopods, which are very sensitive to the stinging cells of the sea anemone. Laboratory observations seem to bear out the instinctive nature of the symbiosis.

The hermit crab signals that it will carry the sea anemone by tapping it with claws and front walking legs. The signal is a specific stroking of a particular intensity which the

Attached. *With its pedal disc now attached to the shell, above, the sea anemone releases the hold its tentacles had on the shell and straightens up.*

Together. *Anemone and crab will stay together as they are below until the crab outgrows the shell. Then they will move to a new one.*

sea anemone will respond to. When it receives the signal, the sea anemone responds by releasing its grip on the substrate and fastening tentacles on the crab's shell. *Calliactis* bends its body into a U-shape, releasing the hold its tentacles have while gripping the crab's shell with its pedal disc. Then it straightens up and rides away atop the hermit crab's shell. When the crab outgrows that shell and moves into a new one, it signals the sea anemone, which moves with it. In other species which have a similar relationship, the crab does not have to stimulate the sea anemone to move. Certain species of sea anemone can take the initiative and transfer their footholds from the old shell to the new one without any action or signal by the crab.

This behavior has been shown to be instinctive. Without the inborn knowledge of the proper signal to the sea anemone, the crabs could not so consistently invite them to attach, nor would the sea anemone shift to new shells with the crabs.

Orienting Toward Light

Animals use their senses both to avoid detrimental environments and to find and remain in favorable conditions. Sea turtles, for instance, just hatched from buried eggs, use their vision to orient themselves and find their way to the sea.

The egg-laying female sea turtles come ashore during their lifetimes solely to dig holes in sandy beaches and deposit their eggs. If the eggs aren't dug up by man or other predators, such as coconut crabs, they hatch in a few weeks. The hatchlings must dig their way out of the sand and then find their way to the sea or perish. If they turn inland, they will probably die of dehydration or predation. If they are not quick enough in locating and getting into the sea, oceanic birds will carry them off and eat them. Even then, if they reach the sea, they may be preyed upon by fish, but their chances of survival are far greater in the water where they can move more easily and quickly. Hatchling sea turtles are believed today to find their way to the sea through visual orientation toward the brightest horizon. And the horizon of the sea is brighter than that over land because the sea reflects light more readily than land. If heavy clouds were to darken the sky at sea while sunny weather prevailed over the land, the young turtles might turn inland.

Hatchling leatherback sea turtles circle briefly after digging their way out of the sand to orient themselves, then they home in on the sea. To check the theory of visual orientation by the baby sea turtles researchers tested them by decreasing or eliminating the light to one eye with light filters and blindfolds. When one eye was completely blindfolded, the turtle circled throughout the ten-minute test period.

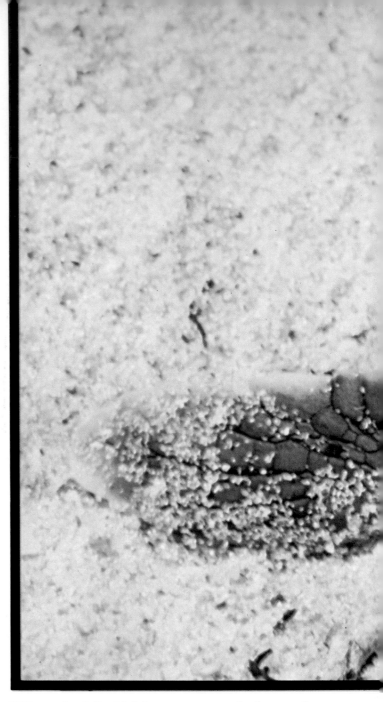

When the blindfold was removed, the turtle immediately took a seaward course. The experimenters also used filters of varying density over one eye of the turtles. They found that the denser the filter, the longer the turtle circled in the direction of the unhindered eye.

Without the instinctive knowledge that the sky is brightest over the sea, these baby sea turtles would have been unable to survive heavy predation. As it is, they undergo heavy attack with few growing to sexual

maturity. It is a disheartening fact that on the island of Europa those hatchlings that venture out of the sand during daytime are almost all exterminated, while the baby turtles that emerge during the night have a much better chance. How do they find the sea during a dark night?

Their instinctive knowledge is apparently never forgotten by the turtles. When the females have matured and come ashore at night to lay their eggs, they are able to find their way back to the sea. Whether they re-

Loggerhead turtle. The turtle which has just hatched and fought its way up through the sand to the surface must now find its way to the sea before predators or the hot sun kill it. Scientific experiments have revealed that hatchling sea turtles of several species move toward the brightest horizon.

member the course they took to get to the nesting place on the beach or are following some unknown stimulation is not clear. Tests similar to those given the hatchling turtles revealed adult turtles, blindfolded over one eye, also circled toward the unhindered eye.

Strange Partners

One of the strangest fish ever observed under the sea is the pearlfish, a slender, elongate, tailless, and scaleless fish of tropical seas. These small fish, have a symbiotic relationship with sea cucumbers, sea stars, tunicates, clams, and sea urchins. Some consort only with one species of their host organism. Others are less particular. The relationship of the pearlfish to the sea cucumber, however, is the strangest of all. The fish chooses its host while it is still very young, forces entry, and from then on considers the cucumber as its home.

The way a pearlfish enters the sea cucumber varies somewhat with the species of pearlfish, and individuals of the same species behave slightly differently. Some enter head-first, but the usual way is for the fish to wait with its head by the sea cucumber's anus until it senses that the anal muscle is relaxing. Then the fish whips its pointed posterior end around forming a U with its body and in-

serts the posterior end into the opening. As it slowly eases its body in, the pearlfish straightens out, and disappears inside, head last. Even very young pearlfish, barely past the second and final larval stage, have been seen entering the bodies of living animals. A few have been found encased in pearl nacre inside clams. Knowledge that they will find shelter in the species their ancestors have always used for protection is almost certainly instinctive. Their ability to get inside their host is probably an inherited skill.

Pearlfish. Backing into the shelter of a sea cucumber's cavity, a pearlfish straightens its body out just before disappearing inside. The fish will eat parts of gills and other organs which the sea cucumber will regrow.

Once inside the sea cucumber, the pearlfish feeds on the branchial gills and reproductive organs of its host, prompting scientists to classify it as a parasite. Because sea cucumbers regenerate virtually every part of their bodies, no great harm is done to it by the eating habits of the pearlfish.

Chapter IX. Learning by Instinct

Instinct and learned behavior are often erroneously thought of as being opposites. Instinct, we have seen, is inherited, inborn knowledge. Learning is acquired after birth. The ability to learn is inherited, however, and is dependent on genetic characteristics.

Animals are limited in their capacity to learn. Some can learn to respond to one type of situation very easily, while another set of conditions may never be mastered. This capacity to learn is definitely an inborn ability that has been programmed into an animal. As a result, instinct and intelligence are not opposite or even separate but complementary, relating and overlapping in varying degrees depending on the species.

In most animals there is an innate predisposition to learn certain things, whether it be the recognition of prey or the recognition of a type of habitat. Such limitations exist in the learning potential of all species. The subtle variations in sounds emitted by dolphins may never be comprehended by a human listening to them because it is not within the capability of the human brain to differentiate between the sounds. There is a very specialized type of learning that exists in animals which imposes even greater limitations on the individual. It is called imprinting and makes the animal receptive to learning a specific response to a certain object during a very limited period of time. If learning does not take place during that time, the individual will never be able to recognize the object or perform the task. The sensitive period is usually early in life and is very limited in regard to the stimulus and response. The response may not occur until adulthood, when a mate is recognized.

The critical learning period is marked in fish that migrate to spawn. Young salmon learn the scent of the home stream in their first year of life. Then they go to sea to stay until the remembered scent draws them back to spawn in the place of their birth.

Just after hatching, baby ducks respond to any large moving object and follow it there-

---

"Instinct and intelligence are not opposite or separate but overlap to some degree."

---

after as though it were the mother, even though it is a person or a box being dragged. The duck was programmed to respond to such a stimulus and after one exposure learned that the object was Mom and that it should be followed. In one species of freshwater fish, the cichlid, the parents do not eat their own young but will consume the offspring of another species of cichlid. They have learned to differentiate their progeny from others by imprinting. Just after hatching, the parents identify those baby fish nearby as their own and from that time on will not eat them. Even when eggs of a different species are put in the cichlid's nest to hatch, the parents will not eat them because they are imprinted on the little fish from their nest, whatever they look like.

Some forms of learning are based directly on instinct even though they are acquired after birth and are truly learned behaviors. Other types of learned behavior, such as marine mammals' ability to use intellectual capacity, will be considered in the next chapter.

Blenny. *Living in tide pools and along the shores of tropical seas, blennies are bottom dwellers who instinctively stay close to and defend their homes in the rocks or retreat to safety.*

Selective Learning

Even the bristling and spiny sea urchin is preyed upon by a few animals. Its protective armor and weaponry are not impenetrable.

Off Point Loma, California, scientists have observed large populations of the purple sea urchin being preyed upon by leather sea stars. Yet farther north, along the coast of Washington State, the two species live side by side, without such predation. Observers found the leather stars off Point Loma pursuing, capturing, and consuming sea urchins. If the sea urchins sensed the presence of the leather stars by touch, they would depress their spines, present their poison-bearing pedicellariae to pinch the offending sea stars, and move away on their tube feet. By taking these actions, the sea urchins showed they were aware of the predator-prey relationship.

It is possible that the sea urchin learned that leather stars could be predators. Unless the sea stars made overt moves toward them, however, there was neither defense nor flight response in the sea urchin. For this reason sea urchins in Washington waters never showed these responses. Why do the Point Loma leather stars prey on the sea urchins while those off the Washington coast do not? Availability of prey is a partial answer. Leather stars prefer a sessile prey, that is, one that does not move. Their first food preference is mussels. Off Point Loma, however, few mussels are available and many sea urchins are. The sea stars there, through associative learning, found sea urchins an acceptable and available second choice, while those in Washington waters have no need of a second choice. Not only did the sea star learn to prey on the sea urchin, but the sea urchin learned to avoid the sea star in the Point Loma waters.

Each learned about the other because it had the innate ability to learn. The learning took place only because there was the ability to respond somewhere in each animal's inherited memory. The ability was inborn. The act was learned.

There is an interesting ecological sidelight to the developing sea urchin-leather sea star story. After the population of sea otters was eliminated in southern California waters and massive increases in pollution occurred, sea urchin populations rose while most other species became unable to reproduce effectively. Will a pollution-tolerant predator eventually alter its feeding behavior to include this new resource of food in its diet?

Sea stars and sea urchins. An ochre sea star (left) does not prey on urchins and thus evokes no avoidance response from them.

Sea urchins. Without the predatory sea star to check their population, the purple sea urchins (right) might upset the ecological balance of the area.

Unnatural Boundaries

A visit to any of the great aquariums of the world reveals the same families of fish on display. Damselfish, wrasses, snappers, grunts, salmon, and morays are commonly seen in the tanks. There are rarely any tuna, requiem sharks, swordfish, marlin, or other species that live in the open sea.

One of the reasons aquarists shun open-water fish species is that they are limited by their innate capacity to learn. These animals bump into the aquarium glass and they frequently injure themselves. They are not able to adapt to confined spaces. In their natural environment they have no problems with impediments. Thus not needing the ability to adapt to such problems, they lack it.

Jacks. Large tanks are needed to keep open-water fish like the jacks in aquariums. Unused to confinement they have difficulty adapting to tight spaces.

Goby. *The frillfin goby can learn to find its own home tide pool, leaping from one to another. The tiny fish can remember its way home for weeks.*

Homing Ability

The learning ability of several species of gobies enables them to find their way from tide pool to tide pool in the coastal waters. These little fish learn the position of their home pools and can return to them whenever the water is deep enough to navigate.

This behavior was first noticed in the frillfin goby, but it appears that a number of other goby species may share it. Experiments have shown that the goby has the inherent ability to learn and remember routes to and from its home pool and can return even in darkness. One animal was able to locate its home after 40 days away from it.

Chapter X. Intelligence and the Birth of Consciousness

Intelligence begins with the ability to learn from experience. Since most animals can learn from experience, we might be tempted to say they all have some degree of intelligence. Learning varies widely from the elementary, one-celled protozoan, which can learn to shun chemicals, to man, whom we place at the apex of intellectual ability.

Intelligence, however, is more than just the ability to learn. It includes finding the relationship between cause and effect, the capacity to correlate various pieces of information previously acquired and stored in memory, and the ability to respond quickly and successfully to new situations. The ability an animal develops to meet all the complicated needs imposed by diverse and constantly changing environments is perhaps at the origin of consciousness.

As consciousness develops in an animal, much of its programmed behavior is lost. In man such adaptation has reached its climax. Instead of passing genetically from one generation to the next information that quickly becomes obsolete, man transmits consciously selected information through tra-

---

"Intelligence is more than simply the ability to learn. It also includes capacity to use acquired knowledge to solve problems never met before."

---

dition and education. One advantage is that our behavior can be modified quickly to meet rapidly changing situations. A disadvantage is that the young of many "intelligent" species are helpless at birth and must undergo extensive education to learn what is needed to survive. It would take perhaps

ten years or more from birth for a human to become physically and mentally competent to survive in wilderness. And it may take a human 25 years or more to learn the skills necessary to become competent in certain professions.

Some young mammals and birds (and probably many more that we have not yet studied) learn the skills needed for survival through play. In such animals learning the necessary skills is acquired and refined by trial and error or, even better, through parental instruction.

Man has evolved as a generalist and is able to adapt to almost any environment. Thus he has successfully lived in some of the harshest habitats—deserts, flood plains, excessively hot or cold areas. It is man's conscious effort to adapt to these varying situations, his intelligence, and his capacity to develop tools that permits him to adjust quickly to changing environments and extreme climatic conditions.

Konrad Lorenz, the father of modern animal behavior studies, once said: "If a man were asked to perform three tasks—to march 35 kilometers [22 miles] in one day, to climb a five-meter [16-foot] rope, and to swim a distance of 15 meters [48 feet] in four-meter [13-foot] depth and pick up a number of objects in a certain order from the bottom, all are activities which a highly nonathletic person like myself can do without difficulty, and there is no single other animal which can duplicate this feat."

Orca. These engaging mammals have large, complex brains which compare favorably to those of men. The orca may indeed be the most intelligent animal on earth.

The Intelligent Octopus

Octopods are considered to be among the most intelligent invertebrates. Their capacity to learn appears to be considerable. Just how much they can learn and how they learn are subjects of great interest to scientists investigating the mechanics of learning.

Octopods have long-term and short-term memories as vertebrates have. The long-term memories are those that are retained for the lifetime of the animal, while short-term memories are those used to solve immediate problems that are not likely to arise again. Humans, for example, learn multiplication tables and usually never forget them. But we often forget a telephone number as soon as we dial it and must look it up again if we want to dial it again a few hours later. Octopods differ from higher animals though, in having two separate learning systems.

One of these is based on visual stimuli, the other on tactile stimuli. Neither is dependent on the other in any way, and if the ability to use one system is blocked or taken away, it has no effect on learning through the other system.

One of the principal ways in which octopods' nervous systems differ from that of vertebrates is in the topography of the brain and its consequent effect on learning and response. In vertebrates, the centers of the brain that specialize in learning are so closely linked with those specializing in movement that if a portion of a learning center is removed, physical responses may be impaired or eliminated. In octopods, however, an entire learning center of the brain can be removed with no effect on the octopus's mobility. Octopods solve problems through experience and actual trial-and-error experiments. Once it has solved a problem, an octopus remembers and can solve the same type of problem in the same way but faster each time. Unfortunately the octopus lives only two or three years, which is very short to take full advantage of its built-in capabilities. The ability of an octopus to learn and to remember has often been demonstrated. *Calypso* divers have observed many octopods building their homes, one of them wedged up a large flat stone with a brick that it had dragged along from quite a distance.

Alert octopus, *right, guards its underwater home. Often an octopus will drag a suitable stone for a long distance to embellish its house.*

Man and octopus, *below, study each other through an aquarium wall. They are en route to Monaco's Oceanographic Museum, where the octopus's intelligence will be tested.*

Tool-Using Sea Otter

The only sea mammal known to use tools consistently in obtaining food is the sea otter. Its paws can almost be used as hands, while the dophin's flippers cannot. The sea otter's intelligence has helped it survive heavy predation by fur traders of the past and commercial abalone fishermen of the present who shoot them because they believe the sea otter is depleting the abalone population. When a colony of sea otters was discovered off the coast of California after a long absence, concerned humans fought to protect them against poachers.

As evidence of its intelligence, the sea otter has been observed using tools in several ways. Diver-photographer Ron Church planted a rock with an abalone firmly gripping it on the bottom as bait. A wily sea otter simply carried the whole rock away. Church planted another abalone on a much larger rock which had a number of smaller rocks around it. When the sea otter came for this one, it picked up one of the smaller rocks and batted at the abalone to knock it off its firmly gripped perch. The rock it used was too small so it picked up another, larger rock which also was too small. After trying three more times, each time with a larger rock than before, the sea otter jarred the abalone loose and made off with it. The sequence of events demonstrated that the sea otter could differentiate the sizes of rocks and that it knew the larger rocks when used with equal force would produce a harder blow against the abalone. Sea otters are also known to use stones, either as hammers or as anvils, to crack shellfish or sea urchins.

Sea otter. *To a sea otter, sea urchins are tasty morsels. The wily mammals use their bodies as levers in prying rocks apart to get to sea urchins.*

Mammals in Captivity

Cetaceans are renowned for their high degree of intelligence which has been observed many times in whales, dolphins, and porpoises at sea or in captivity. The case of a captive pilot whale raises the question of whether mental disorders are one of the prices of higher intelligence when natural bonds are severed. Bimbo, a Pacific pilot whale, was captured by a collecting crew off the California coast for Marineland of the Pacific. Bimbo was the first pilot whale held in captivity for any significant length of time. He underwent training with several species of dolphins and porpoises at the big oceanarium and soon was performing happily and willingly before a large crowd of people every day. Between shows, he swam in his 60-foot-diameter, 22-foot-deep tank as he wished. He shared his prison with a Pacific white-sided dolphin who was a good companion, and eventually a female pilot whale joined him in the big tank. Trainers and management hoped they could record the first live birth of a pilot whale in captivity, but something went wrong. Not only didn't Bimbo sire a pup, but both his tank mates died within a short time of each other. Judging from his subsequent actions, Bimbo was emotionally shattered by the double loss. In each case, holding the dead companion by a flipper, he swam around and around the tank with the body. Several days passed before he would permit his trainers to remove the corpse. Then Bimbo went off his feed. His food intake dropped from 150 pounds of fish a day to nothing. He had to be force-fed distilled water to prevent dehydration. Vitamin injections were pumped into him. Tranquilizers were prescribed and administered by the gram instead of the milligram dosage given to humans. After many inducements, he resumed eating voluntarily, but only about half his former intake. In the

meantime, he had lost close to 1000 pounds, dropping more than 20 percent of his 4500-pound bulk.

Dr. M. E. Webber, a physician who is interested in whales, suggested that Bimbo had become psychoneurotic — a manic-depressive in human psychiatric terms. Bimbo furthered this belief one day as the usual crowd of hundreds peered at him through the glass of his tank. He swam with all his tremendous power against one of the glass ports, shattering it. Thousands of gallons of seawater cascaded out of the tank, sweeping

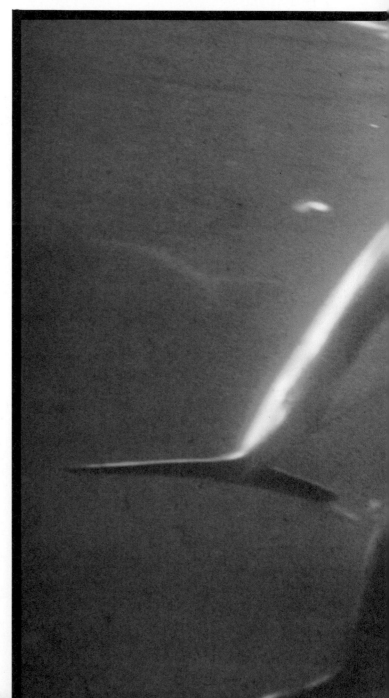

some of the spectators off their feet and drenching many of them as they fled for higher ground.

A few months later, in deference to Bimbo's mental state, and after considerable preparation to be sure he could cope with it, he was released in the sea near a pod of other pilot whales which he is reported to have joined before they all swam away.

Bimbo, the emotional convict returned to freedom, has never been seen since. No friendly pilot whale has showed up on any-one's dock in southern California. Has Bimbo become integrated with the pod or has he been rejected only to succumb to neurasthenia? No one will know for sure. But wardens of wildlife parks and behaviorists know that animals that have been captive for even a short time are rarely accepted by their kind.

Bimbo and companion. Pacific pilot whale and white-sided dolphin swim together in captivity. Mourning inconsolably after the death of tank mates, Bimbo was set free.

Circus Playboys

We have seen that the ability to learn is a necessary but insufficient condition for an animal to be considered intelligent. Some animals have a large capacity to learn about life in the wilderness but refuse to submit to a trainer. Thus brilliant circus animals are not to be considered smarter than more restive individuals.

Seals, especially fur seals and sea lions, are able to learn a variety of tricks and are read-ily trainable. In nature as well as in captivity they like to play. But their intelligence level remains average, probably slightly inferior to that of chimpanzees.

Craig Kitchen has trained a young female sea lion to respond differently to 32 distinct hand signals. Kitchen was curious to learn if the sea lion would perform some of these same behaviors in response to a different stimulus. He therefore taught it to respond to seven new symbols printed on a large card. As a third stimulus, he used tones of

different pitches. Kitchen found that the sea lion did not become confused but was capable of learning and retaining the same responses to the three different stimuli. The same sea lion solved a problem by itself. In its playtime, it often carried a large ball underwater and released it. Usually the ball would pop to the surface, but occasionally it would become trapped in the overhang of an artificial island in the enclosure. The ball was too big for the animal to grasp in its mouth, so it would hold the ball between its front flippers and exhale, causing its body to sink. Still holding the ball, the sea lion would then swim out from under the island and release it. The sea lion did this as often as the ball became trapped under the structure.

In the open sea, fur seals and sea lions display graceful engagement behaviors. They also play with such weaker creatures as marine iguanas without harming them. But they can be mercilessly cruel with such helpless animals as the ocean sunfish.

The Slave Trade

Formidable porpoises—that is really what the so-called killer whales, or gladiator orcas, are. The big males are 25 to 30 feet long and have dorsal fins reaching almost four feet in height. They are powerful swimmers and are beautifully streamlined. Their mouths are big, their teeth strong; they have an accurate sonar echolocation system and a very large brain. They live in herds of 8 to 14 individuals—traveling families, in fact, with a single fully developed male, two to four females, and a varying number of young belonging to several generations. The animals are interdependent.

Until 1964, orcas were known only by their fierce reputations. Then one was captured and kept alive for 87 days. A second orca was accidentally trapped in a fishnet a short time later, and because public interest in killer whales had been stimulated, this one was bought by promoters and put on public display. It lived long enough in captivity to be tamed and ridden like a horse by daring divers. And so a circus star was born.

Commercial fishermen caught orcas by the score. They were flown to aquariums in many parts of the world. It is commonplace today to see these magnificent animals jumping high out of a pool on command, or submitting to having their teeth brushed before a childish audience.

Early whalers reported savage attacks on large baleen whales by groups of orcas that behaved in the same organized way as a pack of wolves. In fact this only happened when the big whales were wounded or killed. But orcas are indeed ingeniously tricky carnivores. Three or four of them have been seen joining forces to overturn an ice floe to catch a seal. Years ago, whale fishermen on the southeastern coast of Australia

reported that orcas made conscious efforts to help them locate pods of humpback whales. While most of the orcas circled the humpbacks, keeping them in a tight group, a few swam toward shore grunting loudly to alert fishermen. The tongues of the slaughtered whales were the reward the orcas received for their services.

Orcas in captivity have given scientists a poor opportunity to study either their physiology or their intelligence: we suspect the validity of studies of animals under shock

kept in shallow tanks. How can we learn about the orca's diving ability when in the open ocean it can dive below a thousand feet? And how can we begin to understand its complex behavior in a tank where echolocation is confused by echoes on the walls?

The study of orcas is only fruitful in the open sea, and it is difficult and time-consuming. Each time we have followed a family of orcas with the *Calypso* in combination with motor launches and a helicopter, we have observed the determination of the dominant

Orca roundup. *Their popularity in marine parks led to the capture of many orcas in Puget Sound where large groups congregated.*

male to lure us away from his family even at the risk of his own life. His maneuvers demonstrated imagination and with his complicated whistling language he was able to direct the route and behavior of his relatives by remote control when they were as distant as a mile from him. Many other observations suggest a highly developed intellect and extended capability for communication.

Yielding dolphin. A paternalistic affection develops between a trainer and captive dolphin. The reverse might happen in the open sea.

Captive Dolphins

When an exuberant, highly social dolphin has been put through the agony of being captured in a throw net, separated from its family, hoisted brutally out of the sea, confined in a training pen, injected with vitamins and antibiotics, and submitted to weeks of brainwashing sessions, it is turned from a proud raider of the sea into a submissive beggar and clown. It is this now perverted creature that some behaviorists attempt to analyze.

In captivity a dolphin is a caricature of itself although it has a tremendous ability to learn and to brilliantly execute complex circus

Free but friendly. *A U.S. Navy-trained dolphin named Dolly is now free and self-sufficient. She frequently visits with humans and plays with children.*

tricks. The dolphin's need for affection has been turned into a comedian's appreciation of applause. The dolphin was taught by its mother to catch live fish for food; now a trainer teaches it to consider dead fish a reward. Calling this "overriding learned behavior" or saying that "Pavlovian response" is used for training is an outrage.

To consider the behavior of a captive dolphin as normal dolphin behavior is tantamount to considering as human behavior the action of those unfortunate victims of Nazi concentration camps who learned to lick the floor at the sound of a whistle for a piece of bread. Behavior forced upon intelligent prisoners is perverted behavior.

Smarter than Man?

Conventional scientists state that it is impossible to compare the intelligence of unrelated species. It may be that a possibility of finding a high intelligence in creatures other than himself is one that man is very reluctant to face. The confusion starts with our self-evaluation. How much of our "intelligence" is innate and how much is borrowed from the heritage of civilization? A human baby lost in the jungle and brought up by wolves does not turn out to be Kipling's Mowgli, but becomes a stupid teenager that cannot be recovered by a tardy education. As long as early mankind relied on oral tradition to accumulate information, civilization lagged. Man's hands have proved even more useful in storing experience (through writing or computers) than in making tools. The "intelligence potential" of our brain can only become effective with the help of our voice, our hands, and our longevity. The toothed whales have brain, voice, longevity, but lack hands. At best they can develop some sort of oral tradition and civilization of a nomadic nature. Before we could "lend our hands" to dolphins, we would have to study them as proud creatures in their natural environment instead of conducting experiments on captive animals that are only neurotic victims whose psyche has been disrupted. When we separate the sociable dolphin from his tribe, we multilate him and he often prefers to die.

Dolly was a female dolphin trained for six years by the U.S. Navy and then released. Having no herd to join she looked for human company and developed a free friendship with Mrs. Jeanne Asbury near Key West. She got her food herself by hunting in the lagoons. She came back to play with the children and her reward was a kiss.

Struggle for Peace

Deep in the infant sea
Inconceivable mornings
Cuddled the stumbling chemistry
Of the first living thing
A stone-hungry cell
A green cell
For the greening of the earth
A cell for peace

Newcomers stole life to feast
The easy loot of a seaweed
Started a deluge of savagery.
As instinct turned into mind
Fangs and armor
Grew in size and might
But giant warriors vanished
Defeated by naked Davids

Thunder from the sky
Awakened the sea
Man's heavy forehead
Rose to gaze at the stars
And soon will grow
A rose that never fades

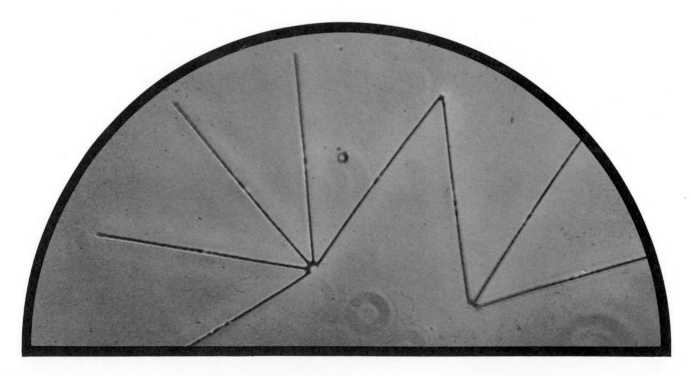

Index

143

ILLUSTRATIONS AND CHARTS:

Sy and Dorothea Barlowe—68-69; Howard Koslow—44-45, 53, 76-77, 105.

PHOTO CREDITS:

John Boland—66, 72-73, 75; Patrick Colin—40-41, 52; Ben Cropp—11, 20-21, 47, 49, 58-59; David Doubilet—78 (bottom), 79 (bottom), 93, 103; Jack Drafahl, Brooks Institute of Photography—36-37, 110-111; Cathy Engel, University of California at Santa Barbara—64 (top); Bill Evans, Naval Undersea Center—132-133; Freelance Photographers Guild: Jerry Jones—106-107, Tom Myers—67, 79 (top), 99, 124, Robert Nansen—2-3, Chuck Nicklin—54-55, 78 (top), 82-83, Dennis Opresko—50-51, Scott Rea—98, Fred Roberts—122, 123, Marvin E. Schneck—26-27, John Stormont—100-101, L. Willinger—14; Henry Genthe—12 (left); Edmond Hobson—17 (top), 19, 25, 108-109; Bob Hollis, Oceanic Films—35; John Hunter, National Marine Fisheries Services, La Jolla, Calif.—91; Hyperion Sewage Plant—142; R. H. Johnson and D. R. Nelson—56-57; Milton Love, University of California at Santa Barbara—61; Richard Mattison—130-131; Richard C. Murphy—12-13 (right), 28, 29, 92; Guy C. Powell—30-31; James Prescott—18; Carl Roessler—65, 74, 80 (middle), 95, 96, 97; D. M. Ross, University of Alberta—32-33, 114-115; John Shoup—39, 42-43 (left), 48, 85; Tom Stack & Associates: A. Bannister (NHPA)—63, Ben Cropp—16, 81, 113, William L. High—127, 136-137, William Stephens—116-117, 118-119, Ron Taylor—5; William M. Stephens—88-89, 90; Paul Tzimoulis—64 (bottom); Don Wobber—22-23.